FATIH AKAY

Top Secret Files

A Journey into the World's 100 Greatest Secrets

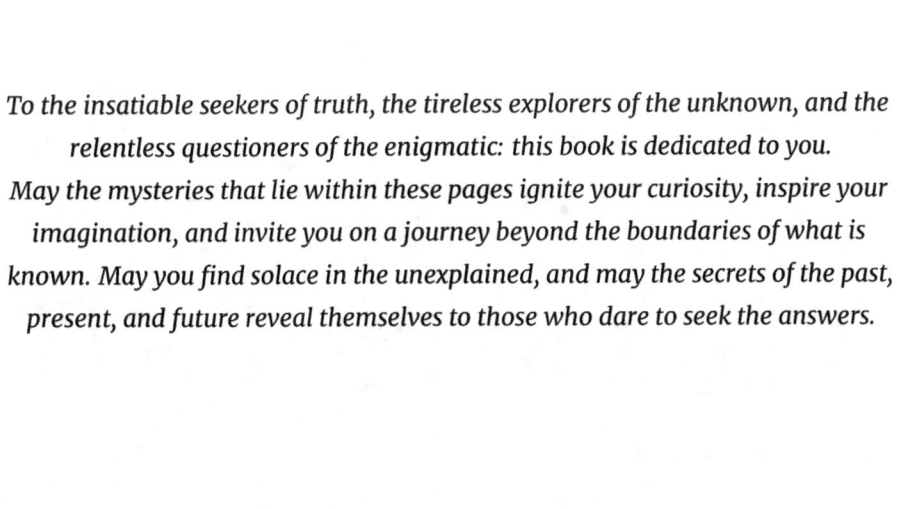

To the insatiable seekers of truth, the tireless explorers of the unknown, and the relentless questioners of the enigmatic: this book is dedicated to you.
May the mysteries that lie within these pages ignite your curiosity, inspire your imagination, and invite you on a journey beyond the boundaries of what is known. May you find solace in the unexplained, and may the secrets of the past, present, and future reveal themselves to those who dare to seek the answers.

"Within the shadows of the unknown lies the wisdom of the ages, waiting to be discovered by those who dare to venture into the depths of mystery."

– Anonymous

Contents

IX Secrets of the Future

X Secrets of Conspiracy Theories

Foreword

In a world where knowledge is both abundant and accessible, it is easy to overlook the mysteries that still elude us. These enigmas, which span across time, space, and human history, captivate our imagination and challenge our understanding of reality. This book brings together a diverse collection of ancient riddles, unexplained phenomena, and enigmatic secrets, inviting readers to embark on a journey into the unknown.

The author has meticulously researched and compiled an astonishing array of mysteries, ranging from the awe-inspiring structures of ancient civilizations to the perplexing riddles of the cosmos. Each chapter delves into the depths of a particular enigma, unraveling the various theories and conjectures that surround it, while also acknowledging the limitations of our understanding. This approach not only encourages the reader to question their assumptions but also fosters a sense of humility in the face of the vast unknown.

As you explore the pages of this book, you will be transported to the hidden chambers of the Great Pyramid, the fabled city of El Dorado, and the enigmatic depths of the Bermuda Triangle. You will ponder the secrets of the universe, the mysteries of the human mind, and the unexplained phenomena that continue to baffle scientists and researchers alike. You will also delve into the world of conspiracy theories, contemplating the powerful forces that may shape our reality from behind the scenes.

It is my hope that this book will inspire a sense of wonder and curiosity in you, as it has in me. May it serve as a reminder that the world is far more complex and intriguing than we often realize, and that the quest for knowledge is an ongoing adventure that knows no bounds. As you turn these pages, let your imagination soar, your skepticism flourish, and your thirst for understanding be unquenchable.

Embrace the mysteries that lie within, and let the journey begin.

Preface

As a lifelong student of history, science, and the mysterious, I have often been captivated by the enigmas that pervade our world. These riddles, which have persisted throughout the ages, serve as a testament to the limits of human understanding and the boundless nature of our curiosity. It is this curiosity that has driven me to compile the collection of mysteries you now hold in your hands.

The purpose of this book is not to provide definitive answers to the questions that have eluded us for centuries, but rather to ignite a passion for discovery and a desire for deeper understanding. Each chapter delves into the heart of a specific mystery, presenting the known facts, the competing theories, and the tantalizing unknowns that continue to confound even the most diligent researchers.

In writing this book, I have been mindful of the need for a balanced approach to the subject matter. While it is tempting to become lost in the world of speculation and conjecture, I have endeavored to ground each exploration in the realm of fact and reason, highlighting the importance of critical thinking and evidence-based inquiry. However, I also encourage readers to keep an open mind and to entertain the possibility that there may be more to these mysteries than meets the eye.

I would like to extend my gratitude to the countless researchers, scientists, historians, and explorers who have dedicated their lives to the pursuit of truth, and who have inspired me to embark on this journey of inquiry. Their tireless efforts have laid the groundwork for this book, and their passion for discovery has been a constant source of motivation throughout the writing process.

As you delve into the pages that follow, I invite you to set aside any preconceived notions you may have and to approach each mystery with a sense of wonder and a spirit of open-minded inquiry. Let the enigmas of the past, the secrets of nature, and the mysteries of the universe serve as a catalyst for your own intellectual and spiritual growth.

It is my sincere hope that this book will not only entertain and inform but also inspire a new generation of seekers to embark on their own journey into the unknown. May you find solace, excitement, and inspiration in these pages, and may the mysteries of our world continue to fuel your passion for discovery.

Happy exploring!

Acknowledgement

This book would not have been possible without the support, encouragement, and contributions of many individuals, to whom I am deeply grateful.

First and foremost, I would like to extend my heartfelt gratitude to my family and friends, who have been an unwavering source of inspiration, motivation, and patience throughout this project. Your belief in me and my passion for the unknown has been instrumental in bringing this book to life.

I am indebted to the numerous researchers, historians, scientists, and explorers whose work has served as the foundation for the mysteries explored within these pages. Your commitment to uncovering the truth and expanding our understanding of the world has paved the way for this compilation.

Finally, to you, the reader: thank you for embarking on this journey with me. Your curiosity, open-mindedness, and desire for understanding are the driving forces that make the exploration of these enigmas worthwhile. It is my hope that the mysteries within these pages spark your imagination and inspire you to continue seeking answers to the unknown.

With immense gratitude and appreciation,

I

Ancient Enigmas

1

The Great Pyramid's Hidden Chambers

Unveiling the Secrets of the Great Pyramid: The Quest for Hidden Chambers

The Great Pyramid of Giza, the last remaining wonder of the ancient world, has captivated the imagination of scholars, archaeologists, and explorers for millennia. Constructed around 2580–2560 BC during the reign of Pharaoh Khufu, this colossal structure continues to defy our understanding of ancient engineering and architectural techniques. One of the most enduring mysteries surrounding the Great Pyramid is the existence of hidden chambers concealed within its massive limestone and granite walls. The search for these elusive spaces has fueled countless expeditions and has given rise to a range of theories, both scientific and speculative.

The Known Chambers

Three main chambers have been discovered within the Great Pyramid: the King's Chamber, the Queen's Chamber, and the unfinished Subterranean Chamber. The King's Chamber, located at the heart of the pyramid, once housed the red granite sarcophagus of Pharaoh Khufu, though his remains have never been found. The Queen's Chamber, situated below the King's Chamber, has been the subject of intense speculation due to its enigmatic construction and the presence of two narrow shafts leading out of the chamber. The Subterranean Chamber, carved deep into the bedrock beneath the pyramid, remains an unfinished and puzzling space, with no clear purpose.

The Search for Hidden Chambers

The quest for hidden chambers within the Great Pyramid dates back to the explorations of Caliph Al-Ma'mun in the 9th century, who, after finding no entrance, ordered a forced tunnel into the pyramid. Since then, generations of explorers have searched for secret spaces using various methods, including traditional excavation and more advanced techniques like ground-penetrating radar and cosmic-ray muon radiography.

In recent years, the ScanPyramids project has utilized cutting-edge technology to search for hidden chambers within the Great Pyramid. In 2016, the project revealed an intriguing anomaly in the form of a thermal hotspot on the eastern side of the pyramid, suggesting a possible void behind the stone. Later, in 2017, the project announced the discovery of a large void above the Grand Gallery, dubbed the "ScanPyramids Big Void." The purpose and contents of this newfound space remain unknown, as accessing it without causing damage to the ancient structure poses a significant challenge.

Theories and Speculation

The existence of hidden chambers within the Great Pyramid has given rise to a variety of theories, ranging from the plausible to the fantastical. Some experts suggest that these chambers may have been built to house the pharaoh's treasures, while others argue that they could be part of an elaborate system of counterweights used during the pyramid's construction. In more speculative circles, theories abound that the hidden chambers may hold ancient records, advanced technology, or even extraterrestrial artifacts.

The search for hidden chambers within the Great Pyramid of Giza remains an ongoing quest that captivates the imagination of both experts and enthusiasts alike. As technology advances and our understanding of this enigmatic structure deepens, it is possible that we may one day uncover the secrets hidden within its walls. Until then, the Great Pyramid will continue to stand as a testament to the ingenuity and ambition of an ancient civilization and a beacon for those who seek to unravel the mysteries of the past.

2

The Lost City of Atlantis

The Lost City of Atlantis: Myth or Reality?

The legend of the lost city of Atlantis has captivated the imagination of scholars, writers, and explorers for centuries. First mentioned by the ancient Greek philosopher Plato in his dialogues "Timaeus" and "Critias," Atlantis was described as a powerful and advanced civilization that ultimately fell out of favor with the gods and was submerged beneath the ocean in a single day and night of catastrophic events. The story of Atlantis has given rise to countless theories, debates, and expeditions seeking to determine the truth behind the myth. Is Atlantis a mere allegory, or does it hold the key to a forgotten chapter of human history?

Plato's Account

According to Plato, Atlantis was a powerful empire that existed around 9,000 years before his own time (circa 360 BC), which would place its origins around 9,600 BC. He described it as a utopian society marked by its advanced technology, skilled navigators, and vast wealth. The island city was said to be larger than Asia and Libya combined and was located "beyond the pillars of Hercules" – generally believed to be the modern-day Strait of Gibraltar. However, when the people of Atlantis became greedy and corrupt, the gods punished them by sinking the entire island beneath the sea.

Theories and Speculations

The story of Atlantis has inspired numerous theories about its possible location and the nature of the civilization itself. Some researchers suggest that the tale is an allegory, created by Plato to illustrate his philosophical ideas about the consequences of hubris and the importance of virtue. Others, however, believe that there may be a kernel of truth in the story, and that Atlantis may have been a real civilization that was lost to time.

Several locations have been proposed as the possible site of Atlantis, including the Mediterranean, the Caribbean, the Azores, and even Antarctica. Some of the more popular theories include the idea that Atlantis was a Minoan civilization on the island of Santorini, which was destroyed by a volcanic eruption, or that it was a highly advanced society located in the Americas that had contact with ancient Egypt and other Mediterranean cultures.

In recent years, some researchers have turned to advanced technology, such as satellite imagery and underwater exploration, in an attempt to find evidence of Atlantis. While these efforts have yet to produce definitive proof, they have uncovered fascinating new information about ancient civilizations and their possible connections to the legend of Atlantis.

Skepticism and Criticism

Despite the many theories and speculations surrounding Atlantis, there is no concrete evidence to support the existence of an advanced civilization that corresponds with Plato's description. Many experts argue that the story is a work of fiction, created by Plato as a cautionary tale to illustrate his philosophical ideas. They point out that no other ancient sources mention Atlantis, and that some of the details in Plato's account, such as the advanced technology and the vast scale of the island, seem implausible given what is known about ancient civilizations.

The legend of the lost city of Atlantis continues to inspire wonder, debate, and exploration. While there is no definitive proof to confirm its existence, the enduring fascination with Atlantis speaks to a deep-rooted human desire to uncover the mysteries of our past and to imagine the possibilities of lost or forgotten civilizations. Whether fact or fiction, the story of Atlantis remains a captivating and enduring enigma that will continue to capture the imagination of generations to come.

3

The True Purpose of Stonehenge

Stonehenge: Decoding the Purpose of an Ancient Monument

Stonehenge, the prehistoric monument located in Wiltshire, England, has fascinated archaeologists, historians, and tourists for centuries. This iconic site, comprised of massive standing stones arranged in a circular formation, was constructed in several stages between 3000 BC and 1600 BC. Despite extensive research and numerous theories, the true purpose of Stonehenge remains one of the world's most enduring archaeological mysteries.

The Construction of Stonehenge

The construction of Stonehenge was a remarkable feat of engineering, considering the limited tools and knowledge available during the Neolithic period. The monument consists of two main types of stones: large sarsen stones, which can weigh up to 25 tons, and smaller bluestones, weighing between 2 to 5 tons each. The sarsen stones are believed to have been sourced from the nearby Marlborough Downs, while the bluestones were transported over 150 miles from the Preseli Hills in Wales. How the builders managed to transport these massive stones and erect them into the iconic formation we see today remains a subject of much debate and speculation.

Theories on the Purpose of Stonehenge

Over the years, a wide range of theories have been proposed regarding the purpose of Stonehenge. Some of the most prominent ideas include:

1. Astronomical Observatory: Stonehenge is aligned with the movements of the sun and the moon, and many researchers believe it may have been used as an ancient astronomical observatory. The monument's entrance aligns with the rising sun on the summer solstice, while the Heel Stone marks the point where the sun sets on the winter solstice. This has led to the idea that Stonehenge was used to track celestial events and mark the changing of the seasons.

2. Sacred Burial Ground: Archaeological excavations have uncovered cremated remains around the site, suggesting that Stonehenge may have been used as a burial ground for high-ranking individuals. Some researchers believe that the monument was a place of ancestor worship, where the dead were honored and remembered.

3. Healing Center: Analysis of the bluestones has revealed that they possess unique acoustic properties, leading some to speculate that Stonehenge was a center for healing. It has been suggested that the stones were believed to have curative powers and that people traveled to Stonehenge seeking relief from illness or injury.

4. Social Gathering Place: Stonehenge may have served as a central location for social gatherings, where people from across the region came together to celebrate, trade, and engage in cultural activities. The construction and maintenance of the monument would have required significant manpower and cooperation, suggesting that it may have played a role in uniting disparate groups.

5. Ritual Site: Many researchers believe that Stonehenge was a site of religious or spiritual significance, where rituals and ceremonies took place. The alignment of the stones with celestial events suggests that these rituals may have been tied to the cycles of the sun and the moon, and the changing of the seasons.

The true purpose of Stonehenge remains an enigma, and it is likely that the monument served multiple functions over the course of its long history. As research continues and new evidence emerges, we may one day gain a clearer understanding of the role this ancient site played in the lives of the people who built it. For now, Stonehenge stands as a testament to the ingenuity, determination, and resilience of our ancestors, a captivating mystery that continues to draw us closer to the past.

4

The Mystery of Easter Island Statues

The Mystery of Easter Island Statues: Unraveling the Secrets of an Isolated Civilization

Easter Island, or Rapa Nui, is a remote volcanic island located in the southeastern Pacific Ocean. It is famous for its nearly 900 monumental statues known as moai, which were carved and erected by the Rapa Nui people between 1250 and 1500 AD. These enigmatic stone figures have long captivated the imagination of researchers and visitors alike, raising questions about the purpose of the statues, the methods used to create and transport them, and the fate of the civilization that built them.

The Moai: Creation and Purpose

The moai were carved from volcanic tuff found in the island's extinct volcano, Rano Raraku. Each statue represents a deceased ancestor, with individual features and characteristics that vary between the figures. The moai were placed on ceremonial platforms called ahu, which were often located along the coastline. It is believed that the statues served as a way to honor and remember the ancestors, and to maintain a spiritual connection between the living and the dead.

The scale and number of the moai are impressive, with some statues reaching up to 33 feet in height and weighing over 80 tons. How the ancient Rapa Nui people managed to carve, transport, and erect these massive statues with their limited tools and resources remains a subject of much debate and speculation.

Transporting the Moai: Theories and Evidence

Researchers have proposed several theories on how the moai were transported from the quarry at Rano Raraku to their final locations on the island. Some of the most popular ideas include:

1. Rolling on Logs: One theory suggests that the moai were moved using logs as rollers, which would have been placed under the statues and then rolled along the ground. However, this method would have required a significant number of logs and manpower, and there is little evidence to support this theory.

2. Walking the Statues: Another theory posits that the Rapa Nui people used ropes to rock the statues from side to side, gradually "walking" them to their destinations. This method is supported by experiments conducted by archaeologists, who have successfully demonstrated that it is possible to move a replica moai using this technique. Additionally, some of the moai have been found in a "face down" position, which could be consistent with the statues falling over during the transportation process.

3. Sledges and Ramps: Another proposal involves the use of wooden sledges to slide the statues across the ground, possibly with the aid of ramps or other structures to help maneuver the statues into place. While this method is technically feasible, there is limited archaeological evidence to support it.

The Decline of the Rapa Nui Civilization

The history of Easter Island also raises questions about the decline of the Rapa Nui civilization. It is believed that the construction of the moai was a significant factor in the island's deforestation and ecological collapse, as trees were cut down to provide logs for transportation and other purposes. This deforestation, coupled with overpopulation and the depletion of natural resources, likely contributed to the decline of the Rapa Nui society.

In addition, the arrival of European explorers in the 18th century brought new diseases, conflicts, and the slave trade, which further devastated the population. By the late 19th century, only a small number of Rapa Nui people remained on the island.

The mystery of Easter Island's statues continues to intrigue researchers and visitors from around the world. While many questions remain unanswered, the moai stand as a testament to the skill, determination, and creativity of the ancient Rapa Nui people. As we continue to study and learn from this unique

5

The Secret of the Nazca Lines

The Secret of the Nazca Lines: Deciphering the Enigma of an Ancient Desert Canvas

The Nazca Lines are a series of ancient geoglyphs located in the Nazca Desert in southern Peru. Created between 500 BC and 500 AD by the Nazca people, these remarkable designs are etched into the surface of the desert, revealing various shapes, animals, plants, and geometric patterns when viewed from above. The Nazca Lines have captivated the imagination of archaeologists, historians, and tourists for decades, and their purpose remains one of the most intriguing unsolved mysteries of the ancient world.

The Creation of the Nazca Lines

The Nazca Lines were created by removing the reddish-brown iron oxide-coated pebbles that cover the surface of the Nazca desert and revealing the light-colored earth underneath. The geoglyphs were made by scraping away the top layer of soil and stones, creating a shallow trench that defines the lines and shapes. This method resulted in the creation of more than 800 straight lines, 300 geometric figures, and 70 biomorphic designs representing animals, birds, and plants.

The scale and precision of the Nazca Lines are extraordinary, with some of the largest figures stretching over 1,200 feet in length. The geoglyphs have been remarkably well-preserved over the centuries, largely due to the arid climate and minimal rainfall in the region.

The Purpose of the Nazca Lines: Theories and Speculation

The true purpose of the Nazca Lines has been the subject of much debate and speculation. Some of the most prominent theories include:

1. Astronomical Alignments: Some researchers believe that the Nazca Lines served as an astronomical observatory, with the geoglyphs aligning to the positions of the sun, moon, and stars. This theory suggests that the Nazca people used the lines to track celestial events and mark important dates, such as solstices and equinoxes. However, the evidence for astronomical alignments is inconclusive, and not all of the lines can be linked to celestial events.

2. Religious or Ritual Significance: Another theory posits that the Nazca Lines had a religious or ritual function, serving as a way for the Nazca people to communicate with their deities or ancestral spirits. The geoglyphs may have been used as sacred pathways for processions, ceremonies, or other religious activities. Some researchers also speculate that the lines were created as offerings to the gods to ensure rain, fertility, or other blessings.

3. Water Resources and Irrigation: Given the arid environment of the Nazca region, some researchers propose that the lines were related to water resources and irrigation systems. The geoglyphs may have been used to mark underground water sources, such as aquifers and subterranean rivers, or to indicate the locations of ancient irrigation canals.

4. Cultural Expression and Art: Another possibility is that the Nazca Lines served as a form of cultural expression or art, representing the beliefs, values, and identity of the Nazca people. The geoglyphs may have been created as a way to share stories, myths, or other aspects of the Nazca culture with future generations.

The enigma of the Nazca Lines continues to captivate researchers and visitors alike. While the true purpose of these ancient geoglyphs may never be fully understood, they remain an incredible testament to the ingenuity, creativity, and resilience of the Nazca civilization. As we continue to study the Nazca Lines and learn more about the people who created them, we gain a deeper appreciation for the rich tapestry of human history and the enduring mysteries that connect us to our past.

6

The Hidden Knowledge of the Mayans

The Hidden Knowledge of the Mayans: Unlocking the Secrets of an Advanced Civilization

Introduction

The Mayan civilization, which thrived from around 2000 BC to 1500 AD in present-day Mexico, Belize, Guatemala, and Honduras, was known for its advanced knowledge in various fields, including astronomy, mathematics, and architecture. Despite the collapse of the Mayan civilization and the destruction of many of their texts by Spanish conquerors, the hidden knowledge of the Mayans continues to intrigue researchers and spark curiosity about their remarkable achievements and the secrets they held.

Astronomy and the Mayan Calendar

The Mayans were exceptional astronomers, meticulously observing and documenting the movements of celestial bodies. They developed a complex and accurate calendar system, which was based on three interlocking cycles: the Tzolk'in (a 260-day sacred calendar), the Haab' (a 365-day solar calendar), and the Long Count (a linear count of days from a set starting point).

The Mayan calendar has been the subject of much fascination, particularly due to its prediction of the end of a significant cycle on December 21, 2012, which some interpreted as the end of the world. Although the Mayans did not predict an apocalypse, their sophisticated understanding of astronomy and the cyclical nature of time remains a testament to their advanced knowledge.

Mathematics and the Concept of Zero

The Mayans were also adept mathematicians, developing a sophisticated number system that included the concept of zero, a mathematical innovation that was independently discovered in only a few ancient civilizations. The Mayan numeral system was vigesimal (base-20), and they used a combination of dots, bars, and special glyphs to represent numbers.

The Mayans applied their mathematical knowledge to various aspects of their society, such as tracking astronomical events, calculating agricultural yields, and designing complex architectural structures.

Architecture and the Mayan City-States

The Mayans were skilled architects and engineers, building magnificent cities with towering pyramids, palaces, temples, and observatories. Some of their most famous structures include the Pyramid of Kukulkan at Chichen Itza, the Temple of the Inscriptions at Palenque, and the Caracol observatory at Chichen Itza.

These architectural marvels display not only the aesthetic mastery of the Mayans but also their understanding of acoustics, geometry, and the principles of load-bearing construction. The Mayan cities were designed to harmonize with their natural surroundings and often incorporated elements that reflected their cosmological beliefs.

Writing and the Mayan Codices

The Mayans had a highly advanced writing system, consisting of more than 800 hieroglyphs representing words, syllables, and concepts. They used this script to record historical events, religious texts, and other aspects of their culture in a variety of mediums, including stone inscriptions, ceramics, and bark-paper books called codices.

Unfortunately, most of the Mayan codices were destroyed by Spanish conquistadors during the colonization of the Americas, leaving only four known surviving codices today. The decipherment of the Mayan script in the 20th century has allowed researchers to unlock many secrets of the Mayan civilization, but countless mysteries remain hidden within the lost texts.

The hidden knowledge of the Mayans, preserved in the remnants of their architecture, writings, and cultural artifacts, continues to captivate and inspire researchers and enthusiasts alike. As we delve deeper into the secrets of this enigmatic civilization, we gain a greater appreciation for the extraordinary achievements of the Mayans and their enduring impact on the fields of astronomy, mathematics, architecture, and beyond. The Mayan civilization stands as a testament to the power of human curiosity, ingenuity, and the relentless pursuit of knowledge.

7

The Curse of Tutankhamun's Tomb

The Curse of Tutankhamun's Tomb: Myth or Reality?

The discovery of Tutankhamun's tomb by British archaeologist Howard Carter in 1922 is one of the most significant archaeological finds of the 20th century. The tomb, located in Egypt's Valley of the Kings, contained a wealth of artifacts and treasures, providing invaluable insights into the life and death of the young Pharaoh. However, the sensational find was soon overshadowed by reports of a mysterious curse, which was said to befall those who dared to disturb the Pharaoh's eternal rest.

The Discovery of Tutankhamun's Tomb

Tutankhamun's tomb was found almost entirely intact, with more than 5,000 artifacts, including the iconic gold funerary mask, the solid gold coffin, and a variety of items intended to assist the Pharaoh in his journey to the afterlife. The discovery captivated the world and ignited a renewed fascination with ancient Egypt.

The Curse: Origin and Popularization

The idea of a curse associated with Tutankhamun's tomb began with media reports of a warning inscribed at the entrance, which supposedly read, "Death shall come on swift wings to him who disturbs the peace of the King." However, this inscription has never been conclusively verified and may have been a fabrication by journalists to enhance the story's intrigue.

The belief in the curse gained further momentum following the death of Lord Carnarvon, the financial backer of the expedition, in 1923, just a few months after the tomb's discovery. Carnarvon's sudden death from an infected mosquito bite fueled rumors that he had succumbed to the curse, even though other members of the expedition, including Howard Carter, did not suffer any such ill fate.

Subsequent Deaths and the Curse

Several other deaths connected to the excavation team and those who visited the tomb have been attributed to the curse of Tutankhamun. However, many of these deaths can be explained by natural causes or coincidence, and statistical analysis has not shown an unusually high mortality rate among those involved in the discovery.

Modern Explanations for the Curse

In recent years, alternative explanations for the perceived curse have been proposed. One such theory suggests that exposure to toxic mold or bacteria present within the tomb may have caused respiratory or other health issues for those who entered, leading to their untimely deaths. Another possibility is that some of the artifacts within the tomb contained traces of hazardous substances, such as arsenic, which could have caused harm if inhaled or ingested.

The Role of the Media and Popular Culture

The media played a significant role in perpetuating the myth of the curse, with sensational headlines and stories capturing the public's imagination. The concept of a mummy's curse has since become a popular trope in literature, movies, and other forms of entertainment, further cementing the idea in the collective consciousness.

The curse of Tutankhamun's tomb, while a captivating and enduring legend, is more likely the product of media sensationalism and popular culture than a genuine supernatural phenomenon. Although the deaths of some individuals connected to the tomb may be shrouded in mystery, they can often be explained by natural causes or coincidence. The true legacy of Tutankhamun's tomb lies not in a mythical curse, but in the incredible wealth of knowledge and understanding it has provided about the life, death, and burial practices of one of history's most enigmatic Pharaohs.

8

The Lost Libraries of Alexandria

The Lost Libraries of Alexandria: The Quest to Unravel the Mysteries of a Vanished Treasure

The Great Library of Alexandria, founded in the 3rd century BCE in Alexandria, Egypt, was one of the most significant intellectual and cultural centers of the ancient world. It housed thousands of scrolls and texts, attracting scholars and researchers from across the Mediterranean. However, the Library's eventual destruction and the loss of its vast collection have left many questions unanswered and inspired a continuing fascination with the secrets it may have held.

The Founding and Importance of the Great Library

The Great Library of Alexandria was founded by Ptolemy I Soter, a general under Alexander the Great, who sought to create a center of learning and research that would rival the renowned Athenian Academy. The Library was part of a larger complex called the Mouseion, which included lecture halls, study rooms, and living quarters for scholars.

The Library's collection was extensive, comprising a vast array of subjects such as astronomy, medicine, mathematics, and literature. It is estimated that at its height, the Library housed anywhere from 40,000 to 400,000 scrolls, making it the largest and most comprehensive library of its time.

The Destruction of the Great Library

The exact circumstances and timeline of the Great Library's destruction are uncertain, and various accounts attribute its demise to different events and periods. Some of the most widely cited theories include:

1. Julius Caesar's invasion of Alexandria in 48 BCE, during which a fire is believed to have spread from the harbor to the Library, damaging or destroying a significant portion of its collection.
2. The decree of Roman Emperor Theodosius I in 391 CE, which ordered the closure of pagan temples, including the Mouseion and the Serapeum, another prominent library in Alexandria.
3. The Muslim conquest of Alexandria in 642 CE, which may have resulted in the destruction of the remaining scrolls and texts.

Regardless of the specific events that led to its destruction, the loss of the Great Library and its vast collection remains one of history's most significant cultural tragedies.

The Search for the Lost Knowledge

The disappearance of the Great Library's collection has fueled a persistent curiosity about the knowledge that may have been lost. Scholars and historians have long speculated about the contents of the Library, which may have included works by influential thinkers such as Euclid, Archimedes, and Sappho, as well as many other texts that have since been lost to history.

Some researchers have sought to reconstruct the Library's collection by examining surviving works that reference or quote texts that were once housed in the Library. Additionally, the discovery of the Nag Hammadi Library, a collection of early Christian texts, and the Oxyrhynchus Papyri, an extensive cache of ancient manuscripts, has provided valuable insights into the types of texts that may have been stored at the Great Library.

The Legacy of the Great Library of Alexandria

While the loss of the Great Library of Alexandria remains a tragedy for the historical and intellectual record, its legacy endures as a symbol of the importance of knowledge, learning, and the preservation of cultural heritage. Today, the Bibliotheca Alexandrina, a modern library and cultural center, stands as a tribute to the spirit of the ancient Library, serving as a reminder of the power of human curiosity and the quest for understanding that transcends generations and civilizations.

9

The Enigma of the Voynich Manuscript

The Enigma of the Voynich Manuscript: Unlocking the Secrets of an Inscrutable Text

The Voynich Manuscript, a mysterious and enigmatic book written in an unknown script and language, has confounded scholars, cryptographers, and linguists since its discovery in the early 20th century. The manuscript's unusual illustrations and seemingly indecipherable text have inspired numerous theories and hypotheses about its origins, purpose, and authorship, making it one of the most intriguing puzzles in the history of cryptography.

The Discovery of the Voynich Manuscript

The Voynich Manuscript was brought to light in 1912 when antique book dealer Wilfrid Voynich purchased it from a collection of manuscripts at Villa Mondragone, Italy. The manuscript, which is currently housed at Yale University's Beinecke Rare Book & Manuscript Library, comprises 240 pages of vellum, with many fold-out illustrations and diagrams. Carbon dating of the vellum places the creation of the manuscript in the early 15th century.

The Mysterious Content of the Voynich Manuscript

The most perplexing aspect of the Voynich Manuscript is its text, written in an unknown script that has yet to be deciphered. The writing system consists of approximately 20-30 distinct characters, with some variations, and is written from left to right.

The manuscript's illustrations offer some clues about its content and are divided into several thematic sections:

1. Herbal: Featuring drawings of plants, some recognizable and others fantastical, accompanied by text that appears to describe their properties or uses.
2. Astronomical: Containing circular diagrams reminiscent of celestial maps, with zodiac symbols and other unidentifiable markings.
3. Biological: Depicting human figures, primarily female, in and around pools or basins, possibly representing bathing or therapeutic rituals.
4. Cosmological: Displaying elaborate, interconnected circular diagrams that resemble star charts or celestial mechanisms.
5. Pharmaceutical: Illustrating various containers, possibly for medicinal or alchemical purposes, with accompanying plant parts and text.
6. Recipes: Consisting of short paragraphs of text, perhaps providing instructions for the preparation or use of the substances mentioned in other sections.

Attempts to Decipher the Voynich Manuscript

Despite numerous attempts by scholars, cryptographers, and amateur enthusiasts, the Voynich Manuscript's text has remained resistant to decipherment. Various theories have been proposed regarding its language, ranging from a natural, yet undiscovered language to a constructed or coded language designed to obscure its true meaning. Some researchers have suggested that the manuscript is a hoax or an elaborate piece of art, with no underlying meaning or content.

The Authorship and Purpose of the Voynich Manuscript

The authorship and purpose of the Voynich Manuscript are as enigmatic as its content. Some theories attribute its creation to historical figures such as Roger Bacon, a 13th-century English philosopher and scientist, or John Dee, a 16th-century mathematician and occultist. Others speculate that it may have been the work of a secretive alchemical or hermetic society.

The purpose of the manuscript is also a subject of much debate, with hypotheses ranging from a medieval herbal or pharmaceutical manual to a treatise on alchemy, astrology, or mysticism.

The Voynich Manuscript remains one of the most baffling and fascinating enigmas in the history of cryptography and the written word. Its indecipher-able text and enigmatic illustrations continue to challenge and captivate researchers, who are eager to unlock the secrets it may hold. Whether the Voynich Manuscript ultimately reveals profound knowledge or simply serves as a testament to the enduring power of human curiosity and the allure of the unknown,

10

The Riddle of the Antikythera Mechanism

The Riddle of the Antikythera Mechanism: Unraveling the Mysteries of an Ancient Astronomical Computer

The Antikythera Mechanism, an ancient Greek artifact discovered more than a century ago, has intrigued and mystified archaeologists, historians, and scientists alike. This complex, clockwork-like device has been hailed as the world's first known analog computer, designed to predict astronomical positions and eclipses. The Mechanism's intricate design and advanced engineering have challenged our understanding of ancient technology and posed numerous questions about its origin, purpose, and the civilization that created it.

The Discovery of the Antikythera Mechanism

The Antikythera Mechanism was discovered in 1901 in the wreckage of an ancient ship off the coast of the Greek island of Antikythera. The artifact, which was initially overlooked due to its corroded and encrusted appearance, was later identified as a highly sophisticated mechanism, consisting of a series of gears, dials, and pointers, all enclosed within a wooden case.

Dating back to around 150-100 BCE, the Antikythera Mechanism is a testament to the extraordinary technical skill and scientific knowledge of the ancient Greeks. Its level of complexity would not be seen again in mechanical devices until the development of European clockwork in the 14th century.

The Function and Purpose of the Antikythera Mechanism

The Antikythera Mechanism was designed to model and predict the movements of celestial bodies, such as the sun, moon, and planets, as well as to track the cycles of the ancient Greek calendar and predict solar and lunar eclipses. The device's intricate gear system allowed it to make highly accurate calculations based on the relative motions of these celestial bodies, following the principles of Greek astronomy and mathematics.

While the primary purpose of the Antikythera Mechanism appears to have been astronomical, its advanced capabilities suggest that it may have also been used for astrological or even navigational purposes. The device's ability to predict the positions of the planets, as well as the timing of eclipses, would have been of great interest to ancient scholars, who believed that celestial events influenced human affairs and could be used to guide decision-making and planning.

The Enigma of the Antikythera Mechanism's Origins

The origin of the Antikythera Mechanism remains one of its most intriguing mysteries. Although its design and function are rooted in Greek astronomy and mathematics, the identity of its creator and the workshop where it was made remain unknown.

Some researchers have proposed that the Mechanism may have been created by or under the supervision of renowned ancient Greek scientists, such as Archimedes, Hipparchus, or Posidonius. The advanced engineering and mathematical principles employed in the device's construction suggest that its creator must have had access to a wealth of scientific knowledge, as well as a highly skilled team of craftsmen.

The Legacy of the Antikythera Mechanism

The Antikythera Mechanism challenges our understanding of the technological capabilities of the ancient world and offers a glimpse into the sophistication and ingenuity of Greek science and engineering. It serves as a reminder that the pursuit of knowledge and the development of advanced technology have deep roots in human history.

The Antikythera Mechanism is just one example of the incredible technological achievements of ancient civilizations. In this chapter, we will explore other examples of ancient technology that have baffled modern researchers, such as the construction of the pyramids and the precision of ancient astronomical observations. We'll also examine the implications of these ancient technological advancements and what they can teach us about the history of human innovation.

II

Hidden Treasures

11

The Elusive Ark of the Covenant

The Ark of the Covenant, a legendary artifact steeped in mystery, has captivated the imaginations of historians, archaeologists, and religious scholars for centuries. Described in the Hebrew Bible as a sacred golden chest that contains the Ten Commandments, the Ark has become the subject of countless myths, theories, and searches. Despite the extensive efforts of countless individuals to locate this enigmatic relic, the Ark remains as elusive as ever. This article explores the history, possible locations, and cultural significance of the Ark of the Covenant.

The Biblical Account

According to the Hebrew Bible, the Ark of the Covenant was constructed by the Israelites following God's command to Moses during their exodus from Egypt. The Ark was designed to house the stone tablets inscribed with the Ten Commandments, along with other sacred objects such as Aaron's rod and a jar of manna. The Ark, which was made of acacia wood and overlaid with gold, had two cherubim facing each other with their wings spread out above the mercy seat.

The Ark played a significant role in the religious life of the Israelites, as it was considered the physical manifestation of God's presence on Earth. It was carried by the priests during important religious ceremonies and battles, and it was believed that the Ark's divine power could lead the Israelites to victory.

Possible Locations and Theories

The Ark's disappearance is shrouded in mystery, as it vanished from the historical record after the Babylonian conquest of Jerusalem in 587 BCE. Numerous theories have emerged over the centuries as to its possible whereabouts:

1. Solomon's Temple: Some scholars believe that the Ark was hidden in a secret chamber beneath Solomon's Temple before the Babylonian invasion to protect it from being captured. However, extensive excavations in the area have not yielded any concrete evidence to support this theory.

2. Ethiopia: The Ethiopian Orthodox Church claims that the Ark is housed in the Church of Our Lady Mary of Zion in Axum, Ethiopia. According to their tradition, Menelik I, the son of King Solomon and the Queen of Sheba, brought the Ark to Ethiopia. However, the Church has never allowed any outside examination of the claimed relic.

3. Southern Africa: Another theory suggests that the Ark was transported to southern Africa by the Lemba people, a Bantu-speaking group with Jewish ancestry. The Lemba claim to have carried the Ark in their migrations across the African continent, but no physical evidence has been found to corroborate this claim.

4. Jordan: The Copper Scroll, one of the Dead Sea Scrolls, provides a list of hidden treasures and artifacts, including the Ark. Some researchers believe that the scroll points to a location in Jordan as the resting place of the Ark. However, no conclusive evidence has been found to support this theory.

Cultural Significance

The Ark of the Covenant continues to captivate the imagination of people worldwide, as it has been featured in various forms of media, most notably the 1981 film "Raiders of the Lost Ark." The Ark's enduring mystery has made it a symbol of divine power, religious devotion, and the search for hidden truths. While the Ark remains elusive, its ongoing impact on culture and our collective imagination is undeniable.

The quest to find the Ark of the Covenant remains one of history's most fascinating and enduring mysteries. Despite numerous theories and potential leads, the location of this sacred artifact continues to elude researchers. Whether the Ark still exists or has been lost to the sands of time, its story will continue to inspire curiosity and intrigue for generations to come.

12

The Mystery of Cleopatra's Tomb

Cleopatra VII, the last pharaoh of ancient Egypt, remains an enduring figure in history, known for her intelligence, beauty, and infamous love affairs with Julius Caesar and Mark Antony. While Cleopatra's life has been extensively documented, her final resting place has never been conclusively identified. Despite centuries of speculation and archaeological exploration, the mystery of Cleopatra's tomb remains unsolved. This article delves into the historical context, potential locations, and ongoing fascination with finding the tomb of the enigmatic Cleopatra.

Cleopatra's Death and Burial

Cleopatra died in 30 BCE, following Egypt's defeat by Octavian (later Emperor Augustus), the future founder of the Roman Empire. According to ancient historians, she committed suicide by allowing an asp, an Egyptian cobra, to bite her. After her death, Octavian allowed Cleopatra to be buried alongside her lover, Mark Antony, who had also taken his own life.

The ancient Greek historian Plutarch wrote that Cleopatra and Mark Antony were buried together in a "golden tomb," adorned with precious stones and located near a temple of the Egyptian goddess Isis. However, Plutarch's account provides no specific details about the tomb's location, leaving the door open for centuries of conjecture and investigation.

Possible Locations

Numerous sites have been proposed as the possible location of Cleopatra's tomb, with varying degrees of evidence and credibility:

1. Taposiris Magna: Since 1998, excavations led by archaeologist Kathleen Martinez have been conducted at the temple of Taposiris Magna, located approximately 45 kilometers west of Alexandria. The temple, dedicated to the god Osiris and the goddess Isis, has yielded several significant discoveries, including statues, pottery, and jewelry. However, despite these finds, Cleopatra's tomb has yet to be discovered at the site.

2. Al Minya: A site near Al Minya, around 300 kilometers south of Cairo, was identified by French archaeologist Franck Goddio as a possible location of Cleopatra's tomb. Goddio and his team discovered an ancient city submerged underwater, including a temple dedicated to Isis. While the site has produced many artifacts, Cleopatra's tomb remains elusive.

3. Alexandria: Some scholars believe that Cleopatra's tomb could be located within the city of Alexandria itself, possibly near the royal palace. However, much of ancient Alexandria now lies beneath the modern city or submerged under the Mediterranean Sea, making archaeological exploration difficult and limited.

The Continuing Fascination

The search for Cleopatra's tomb continues to captivate archaeologists and history enthusiasts alike. The enigmatic nature of her life, combined with the allure of discovering an untouched royal tomb, fuels the ongoing fascination with this ancient mystery.

Cleopatra's tomb, if found, would likely provide a wealth of information about the last days of the Ptolemaic dynasty and Egypt's final era of independence before becoming a Roman province. Moreover, it could offer insights into the burial customs of ancient Egypt's elite and the religious beliefs of the time.

The mystery of Cleopatra's tomb endures as one of the great unsolved enigmas of archaeology. As researchers continue to search for clues and potential leads, the possibility of uncovering Cleopatra's final resting place remains an enticing prospect. Whether her tomb is eventually discovered or remains lost to history, Cleopatra's legacy will continue to captivate and inspire for generations to come.

13

The Quest for the Holy Grail

The Holy Grail, an enigmatic artifact shrouded in myth and legend, has inspired countless tales of valor and spiritual quest for centuries. Often depicted as a chalice or a dish, the Holy Grail is believed to have been used by Jesus Christ during the Last Supper and later to have received His blood during the crucifixion. Though its physical existence remains unproven, the Holy Grail continues to captivate the imagination and represents the enduring human desire for spiritual enlightenment and redemption. This article explores the origins, interpretations, and cultural significance of the quest for the Holy Grail.

Origins of the Legend

The origins of the Holy Grail legend can be traced back to medieval Europe, where it emerged as a central theme in Arthurian literature. The first known reference to the Grail appears in the late 12th-century French romance "Perceval, the Story of the Grail" by Chrétien de Troyes. The tale follows the young knight Perceval on his journey to find the elusive Grail and become its guardian. Over time, the Grail legend evolved and expanded, with various authors adding new elements and interpretations to the story.

Theories and Interpretations

There is no definitive description or explanation for the Holy Grail, leading to a multitude of theories and interpretations:

1. A literal object: Some believe that the Holy Grail is a physical artifact, a sacred vessel with miraculous powers, such as the ability to heal the sick or grant eternal life. Numerous relics and artifacts have been claimed to be the true Grail, including the Valencia Chalice in Spain and the Nanteos Cup in Wales. However, no definitive evidence has been found to support these claims.

2. A metaphor for spiritual growth: Others view the Grail as a metaphor for spiritual enlightenment and the pursuit of a deeper connection with the divine. In this interpretation, the quest for the Grail represents an individual's journey toward self-discovery and spiritual growth.

3. A secret bloodline: In recent years, alternative theories have emerged suggesting that the Holy Grail is not an object but rather a symbol of a secret bloodline descended from Jesus Christ and Mary Magdalene. This theory, popularized by books such as "The Holy Blood and the Holy Grail" and "The Da Vinci Code," has generated significant controversy and debate.

Cultural Significance

The quest for the Holy Grail has left an indelible mark on Western literature and culture. It has been featured in numerous works of fiction, from medieval romances to modern novels and films, often serving as a symbol of divine truth, spiritual fulfillment, and the ultimate human aspiration. The Grail's enduring appeal can be attributed to its ability to inspire both spiritual contemplation and the thrill of adventure.

The quest for the Holy Grail remains one of history's most enduring and captivating mysteries. While the Grail's existence and true nature are open to interpretation, its legend continues to resonate with the timeless human desire for spiritual enlightenment, redemption, and the pursuit of the unattainable. As long as the Holy Grail remains shrouded in myth and legend, it will continue to inspire wonder, curiosity, and the quest for the divine.

14

The Legend of the Lost Dutchman's Mine

The Lost Dutchman's Mine, a fabled gold mine hidden deep within Arizona's Superstition Mountains, has lured treasure hunters, adventurers, and fortune seekers for over a century. Named after a German immigrant, Jacob Waltz, the mine's legend has inspired countless expeditions, wild theories, and tragic tales of misfortune. Despite the numerous attempts to locate the elusive mine, its existence and location remain shrouded in mystery. This article delves into the origins, possible explanations, and enduring fascination with the legend of the Lost Dutchman's Mine.

The Tale of Jacob Waltz

The legend of the Lost Dutchman's Mine centers around Jacob Waltz, a German immigrant who arrived in the United States in the 19th century. According to the story, Waltz discovered a rich gold mine in the Superstition Mountains east of Phoenix, Arizona, around the 1870s. The mine's location was kept secret, known only to Waltz and a few close associates. Waltz allegedly shared vague clues about the mine's whereabouts before his death in 1891, sparking a frenzy of treasure hunters seeking to find the lost fortune.

The Search for the Mine

Since the late 19th century, countless expeditions have been launched in search of the Lost Dutchman's Mine. The Superstition Mountains, a rugged and treacherous terrain, have claimed the lives of many fortune seekers who ventured into the wilderness in pursuit of the elusive gold. Despite extensive efforts and various claims of discovery, the Lost Dutchman's Mine remains undiscovered, fueling speculation and intrigue.

Possible Explanations

The legend of the Lost Dutchman's Mine has inspired numerous theories and explanations:

1. A hidden gold mine: Some believe that the Lost Dutchman's Mine is a genuine, undiscovered gold deposit, hidden away in the remote reaches of the Superstition Mountains. Proponents of this theory often point to Waltz's own gold nuggets, which were said to be unusually high in quality, as evidence of the mine's existence.
2. A conglomeration of smaller mines: Another theory posits that the legend of the Lost Dutchman's Mine may have been inspired by several smaller, abandoned mines in the region. Over time, these mines could have been merged into a single, legendary gold mine in the popular imagination.
3. A fabricated tale: Some skeptics argue that the Lost Dutchman's Mine is nothing more than a tall tale, concocted by storytellers or opportunistic treasure hunters seeking fame or fortune. In this view, the mine is simply a captivating story with no basis in reality.

Enduring Fascination

The legend of the Lost Dutchman's Mine continues to captivate treasure hunters and adventure seekers, as it embodies the archetypal human quest for fortune, discovery, and the pursuit of the unknown. The mine's elusive nature, coupled with the rugged beauty of the Superstition Mountains, has made it a symbol of the American Southwest and the enduring allure of hidden treasures.

The legend of the Lost Dutchman's Mine, whether fact or fiction, has become an integral part of American folklore and the mystique of the Superstition Mountains. As long as the mine remains undiscovered, the legend will continue to inspire curiosity, adventure, and the eternal quest for hidden riches.

15

The Fabled City of El Dorado

The fabled city of El Dorado has captured the imagination of explorers and adventurers for centuries. According to legend, El Dorado was a city made entirely of gold, with streets paved with the precious metal and buildings adorned with jewels and other riches.

The search for El Dorado began in the 16th century, when Spanish conquistadors arrived in South America in search of gold and other treasures. The legend of El Dorado was fueled by stories of indigenous tribes who were said to have vast amounts of gold and other precious metals.

The first known reference to El Dorado came in 1535, when Spanish explorer Gonzalo Jiménez de Quesada received a report from an indigenous informant about a city located in the mountains of Colombia that was said to be rich in gold. Quesada and his men spent several years searching for the city, but they were never able to find it.

Over the centuries, many other explorers attempted to find El Dorado, with varying degrees of success. Some explorers believed that El Dorado was an actual city, while others thought that it was a mythical place that represented the ultimate treasure.

In the 19th century, the legend of El Dorado inspired a new wave of explorers and adventurers, including Sir Walter Raleigh, who led several expeditions in search of the fabled city. However, these expeditions were largely unsuccessful, and the true location of El Dorado remained a mystery.

Despite the lack of concrete evidence, the legend of El Dorado has continued to captivate people around the world. The story of the city made of gold has been immortalized in literature and film, and the idea of a lost city filled with riches remains a powerful symbol of adventure and discovery.

In recent years, archaeologists and historians have made new discoveries that shed light on the origins of the El Dorado legend. It is now believed that the story may have originated with the Muisca people of Colombia, who performed a ritual in which their leader was covered in gold dust and offerings were made to a sacred lake.

While the search for the fabled city of El Dorado may never be fully resolved, the legend continues to inspire and intrigue people around the world. Whether it is a real place or simply a symbol of human longing for wealth and adventure, the story of El Dorado remains a powerful and enduring myth.

16

The Pursuit of King John's Lost Treasure

King John is a historical figure who is well-known for his tumultuous reign over England in the 13th century. However, he is also famous for something else – his supposed hidden treasure. According to legend, King John's lost treasure is still waiting to be discovered, buried somewhere in England.

King John was the youngest son of Henry II and Eleanor of Aquitaine, and he ascended to the throne in 1199. His reign was marked by conflict and controversy, as he fought wars with France and Wales and faced opposition from his own barons. In 1215, he was forced to sign the Magna Carta, which limited the powers of the monarchy and established the principle of due process of law.

According to legend, King John amassed a huge fortune during his reign, including a significant amount of gold and silver. However, when he died in 1216, his treasure was said to have been lost. Some reports suggest that it was buried in a secret location in England, while others claim that it was lost at sea.

Over the years, many people have attempted to find King John's lost treasure, using various methods and techniques. In the 19th century, for example, a group of treasure hunters claimed to have found a hidden hoard of gold and silver in a field in Lincolnshire, which they believed was part of King John's treasure.

In recent years, new technology has been used to try and locate King John's lost treasure. In 2015, a team of archaeologists used ground-penetrating radar to search for the treasure in the grounds of a castle in England. However, they were unable to find any evidence of the lost treasure.

Despite the lack of concrete evidence, the legend of King John's lost treasure continues to captivate people around the world. The idea of hidden treasure, with its promise of riches and adventure, has inspired countless stories and legends throughout history.

While it is possible that King John's treasure may one day be discovered, it remains a mystery to this day. Whether it is buried in a secret location or lost forever, the pursuit of King John's lost treasure remains a compelling and enduring story of human curiosity and the enduring power of myth.

17

The Secret of Oak Island's Money Pit

For over 200 years, the legend of Oak Island's Money Pit has captivated treasure hunters and enthusiasts around the world. Located off the coast of Nova Scotia, Canada, Oak Island is said to contain a mysterious underground chamber filled with treasure, traps, and clues.

According to legend, the Money Pit was discovered in 1795 by a group of teenagers who noticed a depression in the ground and began digging. As they dug deeper, they encountered layers of logs and debris, as well as booby traps such as flooding tunnels and rigged explosives.

Over the years, many treasure hunters and adventurers have attempted to solve the mystery of the Money Pit, using various methods and techniques. Some have used drilling equipment to explore the depths of the island, while others have used scuba diving and underwater cameras to search the surrounding waters.

Despite numerous attempts, no one has been able to definitively solve the mystery of the Money Pit. Some believe that the treasure may be connected to the Knights Templar or the Freemasons, while others think that it may be the lost treasure of pirate Captain Kidd.

In recent years, new technology has been used to try and uncover the secrets of Oak Island. In 2015, for example, a team of researchers used ground-penetrating radar to search for the Money Pit, but their findings were inconclusive.

Despite the lack of concrete evidence, the legend of Oak Island's Money Pit continues to fascinate people around the world. The story of hidden treasure, booby traps, and secret codes has inspired countless books, movies, and TV shows, as well as numerous expeditions and treasure hunts.

Whether or not the Money Pit is real or just a myth, the pursuit of Oak Island's treasure remains a compelling story of human curiosity, perseverance, and the enduring power of mystery.

18

The Hunt for Montezuma's Gold

The search for Montezuma's gold is a tale that has captured the imagination of treasure hunters and adventurers for centuries. According to legend, the Aztec emperor Montezuma II amassed a vast fortune in gold and other precious metals during his reign in the early 16th century, before the arrival of the Spanish conquistadors.

After the Spanish conquered the Aztec empire in 1521, many of the treasures of Montezuma were looted and taken back to Spain. However, some reports suggest that a significant portion of the treasure was hidden and never recovered.

Over the years, many treasure hunters have attempted to find Montezuma's lost gold, using various methods and techniques. Some have searched the ruins of Aztec cities and temples, while others have explored the vast wilderness of Central and South America.

One of the most famous expeditions in search of Montezuma's gold was led by a man named Cornelius "Lame Johnny" Vanderbilt in the 19th century. Vanderbilt claimed to have found a hidden cave filled with gold in Mexico, but his discovery was never confirmed.

In recent years, new technology has been used to try and locate Montezuma's lost gold. In 2018, for example, a team of archaeologists used ground-penetrating radar to search for hidden chambers and tunnels beneath the ruins of the Aztec capital, Tenochtitlan.

Despite numerous attempts, the true location of Montezuma's gold remains a mystery. While it is possible that some of the treasure may still be hidden in the vast wilderness of Central and South America, it is also possible that the legend of Montezuma's gold is simply that – a legend.

Regardless of whether or not the treasure is real, the search for Montezuma's gold remains a compelling story of human curiosity and the enduring quest for wealth and adventure.

19

The Enigma of the Beale Ciphers

The Beale Ciphers are a set of three ciphertexts that are said to contain the location of a treasure worth millions of dollars. According to legend, the treasure was buried by a man named Thomas Jefferson Beale in the early 19th century, and the ciphertexts were left as a clue to its location.

The Beale Ciphers were first made public in the late 19th century, when a pamphlet was distributed that contained the ciphers along with a letter that provided some background information. According to the letter, Beale and a group of fellow adventurers had discovered a treasure in the American Southwest in the early 1800s, and had hidden it in a secret location before returning to the East Coast.

The ciphertexts themselves are complex and difficult to decipher, and have been the subject of much speculation and analysis over the years. Despite numerous attempts, only one of the three ciphertexts has ever been deciphered, and even that decipherment is of questionable accuracy.

Many people have attempted to solve the mystery of the Beale Ciphers, using various methods and techniques. Some have focused on cryptanalysis and computer algorithms, while others have tried to decode the ciphers using historical and geographic clues.

Despite the lack of concrete evidence, the legend of the Beale Ciphers continues to fascinate people around the world. The idea of hidden treasure, encoded messages, and a centuries-old mystery has inspired countless books, movies, and TV shows, as well as numerous treasure hunts and expeditions.

Whether or not the treasure of Thomas Jefferson Beale is real or just a myth, the enigma of the Beale Ciphers remains a compelling story of human curiosity and the enduring power of mystery.

20

The Search for Blackbeard's Treasure

Blackbeard, also known as Edward Teach, was a notorious pirate who terrorized the seas off the coast of North America in the early 18th century. According to legend, Blackbeard amassed a vast fortune in treasure, much of which was buried on various islands and hidden locations.

The search for Blackbeard's treasure has been the subject of many expeditions and treasure hunts over the years, with varying degrees of success. Some have focused on exploring the islands and coastline of the Carolinas, where Blackbeard was known to have operated, while others have attempted to find treasure in more exotic locations such as the Caribbean.

One of the most famous expeditions in search of Blackbeard's treasure was led by a man named William Howard in the 1850s. Howard claimed to have found a hidden cache of treasure on an island off the coast of North Carolina, which he believed was connected to Blackbeard. However, his discovery was never confirmed.

In recent years, new technology has been used to try and locate Blackbeard's lost treasure. In 2018, for example, a team of divers used metal detectors and underwater cameras to search for treasure off the coast of North Carolina, where Blackbeard's ship, the Queen Anne's Revenge, was believed to have sunk.

Despite numerous attempts, the true location of Blackbeard's treasure remains a mystery. While it is possible that some of the treasure may still be hidden on various islands and coastlines, it is also possible that much of it was lost or looted over the centuries.

Regardless of whether or not the treasure is ever found, the legend of Blackbeard's treasure remains a compelling story of human curiosity and the enduring quest for wealth and adventure.

III

Unexplained Phenomena

21

The Enigma of the Bermuda Triangle

The Bermuda Triangle, also known as the Devil's Triangle, is a region in the western part of the North Atlantic Ocean, bounded by Miami, Bermuda, and Puerto Rico. The area is known for its reputation as a site of unexplained phenomena, including the disappearance of ships and airplanes under mysterious circumstances. Despite numerous theories and investigations, the enigma of the Bermuda Triangle remains unresolved to this day.

The mystery of the Bermuda Triangle began in the early 20th century when a series of unexplained events occurred in the region. In 1918, the USS Cyclops, a US Navy cargo ship, vanished in the Bermuda Triangle without a trace, along with its 309 crew members. In the decades that followed, there were reports of numerous other incidents in the area, including the disappearance of Flight 19, a group of five US Navy bombers that vanished in December 1945 during a routine training exercise.

Since then, the Bermuda Triangle has been the subject of many theories and investigations aimed at unraveling the mystery of its unexplained phenomena. Some researchers have suggested that the area is home to a powerful magnetic field that can interfere with the navigation systems of ships and airplanes, causing them to become disoriented and crash or sink.

Others have speculated that the region is a site of extraterrestrial activity or even a portal to another dimension. Some have even suggested that the Bermuda Triangle is the site of a lost city or ancient civilization that is responsible for the strange occurrences in the region.

Despite these theories, there is no conclusive evidence to support any of them. Some researchers have pointed out that the number of incidents in the Bermuda Triangle is not statistically significant, and that similar incidents occur in other parts of the world. Others have suggested that many of the incidents in the Bermuda Triangle can be explained by human error, equipment failure, or natural phenomena such as storms or rogue waves.

Despite the lack of concrete evidence, the mystery of the Bermuda Triangle continues to capture the imagination of the public and inspire countless books, movies, and documentaries. Some people are drawn to the mystery and excitement of the unknown, while others remain skeptical and dismissive of the idea that there is anything mysterious or supernatural about the region.

In the end, the enigma of the Bermuda Triangle remains unresolved, and it is likely to remain a topic of fascination and speculation for years to come. Whether the truth behind the mystery will ever be fully revealed remains to be seen, but for now, the Bermuda Triangle remains a mysterious and intriguing part of the world's oceans.

22

The Mystery of the Tunguska Explosion

The Tunguska Explosion was a mysterious event that occurred in the remote Siberian wilderness of Russia on June 30, 1908. The explosion, which was estimated to have been equivalent to the detonation of 10-15 million tons of TNT, flattened an area of over 2,000 square kilometers and was heard over 1,000 kilometers away. The event has since become known as the Tunguska Event, and it remains one of the most mysterious and unexplained events in history.

The explosion was first observed by a group of reindeer herders who were camped near the Tunguska River in Siberia. They reported seeing a bright, bluish-white light in the sky followed by a deafening explosion that knocked them off their feet. The event was so powerful that it destroyed trees, flattened buildings, and created shock waves that could be felt as far away as the UK.

In the years that followed, scientists and researchers began to investigate the Tunguska Event in an effort to understand what had caused such a powerful explosion. Many theories were put forward, including the possibility that the explosion was caused by a comet or asteroid impact, a volcanic eruption, or even a nuclear explosion.

Despite extensive research, no conclusive evidence has ever been found to support any of these theories. Some researchers have suggested that the explosion may have been caused by a fragment of a comet or asteroid that disintegrated in the atmosphere before hitting the ground. Others have suggested that the event was caused by a nuclear explosion, possibly from an experimental weapon that was being developed at the time.

Despite the lack of a definitive answer, the Tunguska Event remains one of the most intriguing and mysterious events in history. Some researchers have suggested that the event may have been a warning from the universe, while others believe that it was a natural disaster that simply cannot be explained.

Today, the Tunguska Event continues to be studied by researchers and scientists from around the world. New technologies, such as satellite imaging and ground-penetrating radar, have helped to shed new light on the event, but many questions remain unanswered. Until a definitive answer is found, the Tunguska Event will remain a fascinating and enigmatic event that continues to capture the imagination of people around the world.

23

The Puzzling Phenomenon of Crop Circles

Crop circles are a phenomenon that has fascinated and puzzled people for decades. These intricate patterns and designs, often found in fields of wheat, corn, and other crops, are created by flattening and bending the stalks of the plants. While some people believe that crop circles are created by extraterrestrial beings, others suggest that they are man-made, created by artists or pranksters. Despite numerous investigations, the true origin and purpose of crop circles remain a mystery.

Crop circles first gained widespread attention in the late 1970s, when a farmer in southern England discovered a strange pattern in his wheat field. The design was a series of circles, with smaller circles within larger circles, arranged in a precise and intricate pattern. The farmer reported his discovery to the local media, and soon, crop circles began to appear in fields all over England and other parts of the world.

Many people were convinced that crop circles were created by aliens, possibly as a form of communication or message to humans. Others were more skeptical, suggesting that the circles were created by hoaxers using simple tools and techniques.

In the years that followed, numerous investigations were conducted to determine the origin and purpose of crop circles. Some researchers suggested that the circles were created by natural phenomena, such as atmospheric conditions or the weight of snow on crops. Others suggested that they were created by animals, such as badgers or deer.

Despite these theories, many people remained convinced that crop circles were created by aliens or other supernatural forces. Some even claimed to have witnessed strange lights or other unexplained phenomena in the vicinity of crop circle sites.

In recent years, however, many crop circle experts and researchers have come to the conclusion that most crop circles are, in fact, man-made. These designs are created using simple tools and techniques, such as ropes, boards, and planks, and are often created as elaborate pranks or works of art.

Despite this revelation, crop circles remain a popular and intriguing phenomenon, and new designs continue to appear in fields around the world. While the true origin and purpose of crop circles may never be fully understood, they continue to capture the imagination of people around the world, inspiring countless theories, investigations, and debates.

24

The Secret of the Marfa Lights

The Marfa Lights are a mysterious phenomenon that has fascinated and puzzled people for over a century. These unexplained lights, which appear in the night sky near Marfa, Texas, have been the subject of countless investigations, theories, and debates. Despite years of research, however, the true origin and nature of the Marfa Lights remain a mystery.

The Marfa Lights were first observed in the late 19th century, when settlers in the area reported seeing strange lights in the sky at night. These lights, which appear as glowing orbs or hovering shapes, seem to move in unusual patterns and colors. While some people have suggested that the Marfa Lights are simply a natural phenomenon, such as swamp gas or ball lightning, others believe that they are the result of extraterrestrial activity or supernatural forces.

Over the years, numerous investigations have been conducted in an effort to determine the true nature of the Marfa Lights. Scientists and researchers have used a variety of tools and techniques, including cameras, radar, and thermal imaging, to try to capture and analyze the lights. Despite these efforts, however, no conclusive evidence has ever been found to explain the phenomenon.

Some theories suggest that the Marfa Lights are caused by the reflection of car headlights or other artificial sources of light. Others suggest that they are the result of natural phenomena, such as temperature inversions or atmospheric conditions. Still, others believe that the lights are the result of supernatural activity, such as ghosts or spirits.

Despite the lack of a definitive explanation, the Marfa Lights continue to capture the imagination of people around the world. Visitors to the area come from far and wide to catch a glimpse of the mysterious lights, and many have reported experiencing strange sensations or feelings while observing them.

In the end, the true origin and nature of the Marfa Lights may never be fully understood. While some people may continue to believe that they are the result of supernatural or extraterrestrial activity, others will continue to search for a more rational explanation. For now, the Marfa Lights remain one of the most intriguing and enigmatic phenomena in the world, inspiring countless theories and investigations.

25

The Taos Hum: An Unidentified Auditory Mystery

The Taos Hum is an unexplained auditory phenomenon that has been experienced by residents of the town of Taos, New Mexico, since the early 1990s. The hum is described as a low-frequency sound that is audible to some people but not others. While many theories have been put forward to explain the Taos Hum, its true origin remains a mystery.

The Taos Hum was first reported in the early 1990s, when a group of residents in Taos began to experience a low-pitched hum that seemed to come from nowhere. The hum was described as a low-frequency sound, similar to the sound of an idling diesel engine or a distant rumble of thunder. While some residents were able to hear the hum, others could not.

Over the years, many investigations have been conducted in an effort to determine the origin and nature of the Taos Hum. Scientists and researchers have used a variety of tools and techniques, including sound recording equipment and surveys of affected residents, to try to capture and analyze the hum. Despite these efforts, however, no conclusive evidence has ever been found to explain the phenomenon.

Many theories have been put forward to explain the Taos Hum. Some researchers suggest that it may be caused by natural phenomena, such as atmospheric pressure changes or the movement of tectonic plates. Others suggest that it may be caused by man-made sources, such as electrical transformers or industrial machinery. Still, others believe that the Taos Hum may be a psychological phenomenon, possibly caused by stress or anxiety.

Despite the lack of a definitive explanation, the Taos Hum continues to be experienced by residents of the town and has even been reported in other parts of the world. Some people are more affected by the hum than others, with some reporting that it causes headaches, dizziness, and other physical symptoms.

In the end, the true origin and nature of the Taos Hum remains a mystery. While some people may continue to believe that it is caused by natural or man-made sources, others will continue to search for a more rational explanation. For now, the Taos Hum remains an enigmatic and intriguing auditory phenomenon that has captured the imagination of people around the world.

26

The Curious Case of the Dancing Plague

The Dancing Plague, also known as the Dancing Mania, was a strange and mysterious phenomenon that occurred in Europe during the late Middle Ages and early Renaissance. The Dancing Plague was characterized by groups of people who suddenly began to dance uncontrollably, often for days or even weeks at a time. The cause of the Dancing Plague remains a mystery, and it has become a fascinating and perplexing part of history.

The Dancing Plague first appeared in the early 14th century, in what is now modern-day Belgium. It quickly spread to other parts of Europe, including Germany and France. The phenomenon was most common in areas that were heavily affected by the Black Death, a deadly pandemic that swept through Europe in the mid-14th century.

During the Dancing Plague, groups of people would suddenly begin to dance uncontrollably in the streets, often for hours or even days at a time. Some people would even dance until they collapsed from exhaustion or died. Many people who were affected by the Dancing Plague reported feeling as though they were possessed by some kind of supernatural force.

Over the years, many theories have been put forward to explain the Dancing Plague. Some researchers suggest that it may have been caused by a contagious disease, such as ergotism, which is caused by a fungus that grows on rye and wheat. Others suggest that it may have been caused by mass hysteria or some kind of psychological disorder.

Despite these theories, the true cause of the Dancing Plague remains a mystery. Some researchers suggest that it may have been caused by a combination of factors, including environmental, psychological, and social factors. Whatever the cause, the Dancing Plague remains one of the most intriguing and enigmatic phenomena in history.

In the end, the Dancing Plague is a reminder of the strange and mysterious aspects of human behavior. While we may never fully understand the true cause of this phenomenon, it continues to capture the imagination of people around the world and remains a fascinating part of history.

27

The Unsolved Phenomenon of Ball Lightning

Ball lightning is a mysterious phenomenon that has been observed by people for centuries. It is described as a glowing, spherical object that appears during thunderstorms and may last for several seconds to a few minutes before disappearing. Despite numerous investigations and scientific studies, the true nature and cause of ball lightning remain a mystery.

Ball lightning was first described in the early 18th century, and reports of the phenomenon have been recorded throughout history. The balls are typically described as being the size of a grapefruit or basketball and appearing in a variety of colors, including white, yellow, orange, and blue. Witnesses often report seeing the balls move erratically and unpredictably, floating through the air or bouncing off objects.

Over the years, many theories have been put forward to explain the cause of ball lightning. Some researchers suggest that it may be a plasma phenomenon, caused by ionization of the air during thunderstorms. Others suggest that it may be caused by electromagnetic radiation or the interaction of lightning with the earth's magnetic field.

Despite numerous investigations, however, no conclusive evidence has ever been found to explain the true nature of ball lightning. While some scientists continue to study the phenomenon, others believe that it may simply be a rare and unexplained natural occurrence.

Despite its mystery, ball lightning remains a fascinating and intriguing phenomenon, inspiring countless theories and investigations. It has been the subject of numerous books, documentaries, and scientific studies, and continues to capture the imagination of people around the world. In the end, the true nature of ball lightning may remain a mystery, but its enigmatic nature serves as a reminder of the mysteries and wonders of the natural world.

28

The Strange World of Spontaneous Human Combustion

Spontaneous Human Combustion (SHC) is a phenomenon that has puzzled and fascinated people for centuries. SHC refers to cases where a human body appears to have caught fire and burned, often leaving behind little or no evidence of external fire sources. While many theories have been put forward to explain the phenomenon, the true cause of SHC remains a mystery.

SHC cases have been reported throughout history, with the earliest recorded case dating back to the 16th century. In these cases, the victim's body is often found with little or no damage to surrounding objects or furniture, suggesting that the fire was isolated to the body itself. In some cases, only certain parts of the body, such as the torso or head, are burned, while other parts remain intact.

Over the years, many theories have been put forward to explain SHC. Some researchers suggest that the phenomenon may be caused by a build-up of gases in the body, which can ignite when exposed to a spark or flame. Others suggest that it may be caused by chemical reactions in the body, or by the body's own electrical fields.

Despite these theories, no conclusive evidence has ever been found to explain the true cause of SHC. While some researchers continue to study the phenomenon, others believe that it may simply be a rare and unexplained natural occurrence.

SHC cases continue to be reported to this day, with many of them receiving widespread media attention. Some people remain convinced that SHC is caused by supernatural or paranormal forces, while others believe that it is a purely natural phenomenon that is simply not well understood.

In the end, the true cause of SHC may remain a mystery. While the phenomenon continues to fascinate and intrigue people around the world, it also serves as a reminder of the strange and mysterious aspects of human existence.

29

The Mystery of Skyquakes

Skyquakes, also known as mystery booms or unexplained explosions, are a phenomenon that has puzzled and frightened people for centuries. Skyquakes refer to loud booming or rumbling sounds that occur in the sky, often without any obvious explanation or source. While many theories have been put forward to explain skyquakes, the true cause of these mysterious sounds remains a mystery.

Skyquakes have been reported throughout history, with some of the earliest recorded cases dating back to the 19th century. In these cases, witnesses describe hearing loud, booming sounds that seem to come from the sky. The sounds are often described as being similar to thunder or artillery fire, but without any visible lightning or smoke.

Over the years, many theories have been put forward to explain skyquakes. Some researchers suggest that the sounds may be caused by natural phenomena, such as atmospheric pressure changes or the movement of tectonic plates. Others suggest that they may be caused by man-made sources, such as sonic booms from aircraft or explosions from military testing.

Despite these theories, no conclusive evidence has ever been found to explain the true cause of skyquakes. While some researchers continue to study the phenomenon, others believe that it may simply be a rare and unexplained natural occurrence.

Skyquakes continue to be reported to this day, with many people reporting hearing the loud, booming sounds. Some people remain convinced that skyquakes are caused by supernatural or extraterrestrial forces, while others believe that they are simply a natural occurrence that is not yet fully understood.

In the end, the true cause of skyquakes may remain a mystery. While the phenomenon continues to fascinate and intrigue people around the world, it also serves as a reminder of the strange and mysterious aspects of our world that are yet to be fully understood.

30

The Unexplained Disappearance of the Flannan Isles Lighthouse Keepers

The disappearance of the Flannan Isles lighthouse keepers is a mysterious and chilling story that has fascinated people for over a century. In December 1900, three lighthouse keepers stationed on the remote island of Eilean Mor in the Flannan Isles off the west coast of Scotland disappeared without a trace, leaving behind a series of puzzling clues that have never been fully explained.

The Flannan Isles lighthouse had been in operation for just over a year when the three keepers, Thomas Marshall, James Ducat, and Donald McArthur, vanished. On December 15, a supply ship arrived at the island to find that the lighthouse was unmanned and the keepers were nowhere to be found. The relief lighthouse keeper, Joseph Moore, discovered that the entrance gate to the compound was locked, the lamps were lit, and the clocks had stopped.

Upon further investigation, Moore found that the beds had not been slept in and that the table had been set for breakfast, suggesting that the keepers had disappeared suddenly and without warning. A search of the island revealed no sign of the missing men, and subsequent investigations turned up no concrete evidence of foul play or accident.

Over the years, many theories have been put forward to explain the disappearance of the Flannan Isles lighthouse keepers. Some suggest that the men were swept away by a sudden storm or rogue wave, while others suggest that they were kidnapped or murdered by passing sailors or pirates. Some even suggest that the men were abducted by extraterrestrial beings.

Despite numerous investigations, however, no conclusive evidence has ever been found to explain the true cause of the disappearance. The mystery of the Flannan Isles lighthouse keepers remains unsolved, and their fate continues to intrigue and baffle people around the world.

In the end, the disappearance of the Flannan Isles lighthouse keepers serves as a reminder of the mysterious and inexplicable aspects of our world. While we may never fully understand what happened to these men, their story continues to captivate our imaginations and inspire countless theories and investigations.

IV

Secrets of the Universe

31

The Origin of the Universe

The Origin of the Universe: Unraveling the Cosmic Mystery"

The origin of the universe has always been a topic of profound interest and fascination for scientists, philosophers, and laypeople alike. Over the years, countless theories have been proposed in an attempt to explain the birth and evolution of our universe. With advances in technology and a deeper understanding of the cosmos, we are inching closer to solving this age-old mystery. In this article, we will delve into the prevailing theories and scientific evidence that seeks to shed light on the enigmatic origin of the universe.

The Big Bang Theory

The most widely accepted explanation for the origin of the universe is the Big Bang Theory. This theory postulates that the universe began as an extremely hot, dense, and compact singularity, which underwent a rapid expansion approximately 13.8 billion years ago. As the universe expanded, it cooled down, allowing the formation of subatomic particles, atoms, and eventually galaxies, stars, and planets.

Support for the Big Bang Theory comes from several lines of evidence, including:

1. Cosmic Microwave Background (CMB) radiation: Discovered in 1964,

the CMB is the afterglow of the early universe, a relic of the intense heat that permeated the cosmos shortly after the Big Bang. This radiation has been observed in all directions of the sky and is remarkably uniform, supporting the notion of a singular origin.

2. The abundance of light elements: The Big Bang Theory predicts that the early universe was dominated by hydrogen, helium, and trace amounts of lithium. Observations of the cosmic elemental composition align with these predictions.

3. Hubble's Law and the expansion of the universe: Astronomer Edwin Hubble discovered that galaxies are moving away from us, and the farther away they are, the faster they are moving. This observation indicates that the universe is expanding, supporting the idea that it began as a single, dense point.

Alternative Theories and the Multiverse Hypothesis

While the Big Bang Theory is the dominant explanation for the origin of the universe, alternative theories have been proposed as well. Some of these include:

1. The Steady State Theory: This now largely discredited theory proposed that the universe has no beginning or end and is infinitely old. It suggested that matter is continuously created to maintain a constant density as the universe expands.

2. The Cyclic Universe Theory: This idea posits that the universe goes through cycles of expansion, contraction, and rebirth, driven by the interplay of dark matter and dark energy.

3. The Multiverse Hypothesis: This speculative concept suggests that our universe may be just one of many universes existing in parallel, each with its own unique laws of physics and cosmic history.

Although the Big Bang Theory is currently the most compelling explanation for the origin of the universe, there is still much to learn about the early cosmos and its subsequent evolution. As scientists continue to gather data and refine their understanding of the universe, we move closer to unlocking the secrets of its enigmatic birth. The origin of the universe remains a tantalizing mystery, one that continues to inspire wonder and curiosity about our place in the cosmos.

10 facts about the origin of the universe:

1. The most widely accepted theory of the origin of the universe is the Big Bang theory, which proposes that the universe began as a hot, dense, and infinitely small singularity about 13.8 billion years ago.
2. The Big Bang theory was first proposed in the 1920s by Belgian astronomer Georges Lemaître.
3. The cosmic microwave background radiation, which is believed to be leftover radiation from the Big Bang, was first discovered in 1964 by American physicists Arno Penzias and Robert Wilson.
4. The universe is currently expanding at an accelerating rate, which was first discovered in the late 1990s through observations of distant supernovae.
5. The early universe was primarily composed of hydrogen and helium, with small amounts of other elements such as lithium and beryllium.
6. The first stars and galaxies began to form about 100 million years after the Big Bang.
7. The Large Hadron Collider (LHC) is a particle accelerator that was built to study the fundamental particles and forces of nature, including the conditions of the early universe.
8. The concept of inflation, which proposes that the universe underwent a brief period of exponential expansion immediately after the Big Bang, was first introduced in the 1980s.
9. The idea of a multiverse, or the existence of multiple parallel universes, is a topic of debate among physicists and cosmologists.
10. There are still many mysteries and unanswered questions about the origin and nature of the universe, including the nature of dark matter and dark energy, which are thought to make up the majority of the

universe's mass and energy.

32

The Mystery of Dark Matter

Dark matter, an enigmatic substance that neither emits nor absorbs light, has perplexed scientists since its existence was first proposed in the 1930s. Despite being invisible, dark matter is thought to make up approximately 27% of the universe's total mass-energy content, dwarfing the 5% attributed to ordinary matter. Its gravitational effects on galaxies and galaxy clusters are the primary evidence for its existence. In this article, we will explore the evidence for dark matter, the leading theories regarding its nature, and the ongoing quest to detect and understand this cosmic conundrum.

The Evidence for Dark Matter

Dark matter was first theorized when Swiss astronomer Fritz Zwicky observed that the mass of galaxy clusters, as calculated from the motion of galaxies within them, was much greater than the visible mass. This discrepancy suggested the presence of an unseen mass, which he dubbed "dark matter." Key pieces of evidence for the existence of dark matter include:

1. Galaxy rotation curves: Observations of the rotation of galaxies indicate that their outer regions move faster than expected, based on the visible mass. This suggests that dark matter provides additional gravitational force, influencing the motion of stars and gas in galaxies.

2. Gravitational lensing: The bending of light by massive objects, known

as gravitational lensing, provides another clue to the presence of dark matter. Observations of lensing effects around galaxy clusters reveal more mass than can be accounted for by visible matter alone.

3. Cosmic Microwave Background (CMB) radiation: The CMB, a relic of the early universe, offers valuable insights into the distribution of matter in the cosmos. Studies of the CMB reveal a consistent pattern of density fluctuations, supporting the existence of a dominant dark matter component.

The Nature of Dark Matter

While the evidence for dark matter is robust, its exact nature remains elusive. Several theories have been proposed to explain the composition of dark matter:

1. WIMPs (Weakly Interacting Massive Particles): WIMPs are hypothetical particles that interact weakly with ordinary matter and radiation. They are leading candidates for dark matter, as they could provide the necessary gravitational influence without being detectable through electromagnetic radiation.

2. Axions: These hypothetical particles are much lighter than WIMPs but are thought to be abundant enough to account for dark matter. Axions are predicted to have unique interactions with electromagnetic fields, which could potentially be exploited for their detection.

3. MACHOs (Massive Astrophysical Compact Halo Objects): MACHOs are large, non-luminous objects, such as black holes or brown dwarfs, that could account for some of the missing mass in the universe. However, they are now considered unlikely to explain the entirety of the dark matter problem.

The Search for Dark Matter

The race is on to directly detect dark matter particles in laboratory experiments. Several experiments, such as the Large Hadron Collider at CERN, are searching for WIMPs, while others, like the Axion Dark Matter Experiment

(ADMX), focus on detecting axions. So far, no definitive detections have been made, but these efforts continue to refine our understanding of dark matter's properties and possible interactions with ordinary matter.

Dark matter remains one of the most tantalizing mysteries in modern astrophysics. As scientists continue to search for direct evidence and develop new theories about its nature, the enigma of dark matter fuels our curiosity and drives us to explore the hidden forces that shape our universe. Unraveling the mystery of dark matter will not only deepen our understanding of the cosmos but may also reveal new insights into the fundamental laws of physics.

10 facts about the mystery of dark matter:

1. Dark matter is a form of matter that does not interact with light or any other form of electromagnetic radiation, making it invisible to telescopes and other instruments.
2. The existence of dark matter was first inferred in the 1930s by Swiss astronomer Fritz Zwicky, who noticed that the observed mass of galaxy clusters did not account for the gravitational forces that held them together.
3. It is estimated that dark matter makes up approximately 27% of the total mass and energy of the universe, while ordinary matter makes up only about 5%.
4. The nature of dark matter is still unknown, but it is thought to be made up of particles that are not part of the Standard Model of particle physics.
5. Many experiments have been conducted to detect dark matter particles, including searches using underground detectors, particle colliders, and observations of cosmic rays.
6. The Large Hadron Collider (LHC) at CERN in Switzerland is one of the most powerful particle colliders in the world and has been used to search for dark matter particles.
7. The most widely accepted theory of dark matter is the cold dark matter

theory, which proposes that dark matter particles move relatively slowly and clump together to form large structures.

8. The Bullet Cluster is a galaxy cluster that provides strong evidence for the existence of dark matter, as it shows that the majority of the mass in the cluster is located in the dark matter halo surrounding the visible matter.

9. The study of dark matter is important for understanding the formation and evolution of galaxies, as well as the large-scale structure of the universe.

10. There are still many unanswered questions about the nature of dark matter, including its properties, composition, and interactions with other forms of matter and energy.

33

The Search for Extraterrestrial Life

The possibility of extraterrestrial life has long captivated the human imagination. Are we alone in the universe, or do other forms of life exist on distant planets and moons? The search for extraterrestrial life seeks to answer this profound question, bridging scientific disciplines such as astronomy, biology, and astrochemistry. In this article, we will explore the methods and tools employed in the hunt for life beyond Earth, the potential habitats for extraterrestrial life, and the implications of such a discovery.

Searching for Signs of Life

The search for extraterrestrial life is multifaceted, utilizing a variety of approaches to detect potential biosignatures – signs of life that may be present in the atmosphere, surface, or subsurface of a celestial body. Some key strategies include:

1. Observing Exoplanets: The discovery of thousands of exoplanets – planets orbiting stars outside our solar system – has revolutionized the search for life. By studying exoplanets' atmospheres and characteristics, scientists can assess their potential habitability and search for chemical signatures indicative of life.

2. Robotic Exploration: Space missions, such as NASA's Mars rovers and the European Space Agency's Rosetta mission to comet 67P/Churyumov-Gerasimenko, have provided invaluable insights

into the potential for life on other celestial bodies. These robotic explorers analyze surface and subsurface materials to search for organic molecules and other signs of past or present life.

3. Radio Astronomy and SETI: The Search for Extraterrestrial Intelligence (SETI) program and other initiatives use radio astronomy to search for signals from extraterrestrial civilizations. By scanning the skies for non-random signals, these efforts hope to detect evidence of intelligent life in the cosmos.

Potential Habitats for Extraterrestrial Life

Scientists have identified several celestial bodies within our solar system and beyond that may harbor the necessary conditions for life:

1. Mars: Evidence of past liquid water and the presence of organic molecules have made Mars a prime target in the search for extrater-restrial life. Subsurface water reserves or past habitable environments may hold clues to past or present microbial life.

2. Icy Moons: Jupiter's moon Europa and Saturn's moon Enceladus are believed to harbor subsurface oceans beneath their icy crusts. The presence of liquid water, geothermal activity, and organic molecules makes these moons potential habitats for life.

3. Exoplanets in the Habitable Zone: Planets within the habitable zone – the region around a star where conditions may be suitable for liquid water – are prime candidates for hosting life. The recent discovery of potentially habitable exoplanets, such as Proxima Centauri b and TRAPPIST-1e, has heightened interest in these distant worlds.

Implications of Discovering Extraterrestrial Life

The discovery of extraterrestrial life would have profound implications for our understanding of the universe and our place in it. Such a finding could:

1. Inform our understanding of the origins and distribution of life in the cosmos, and potentially offer insights into the processes that led to the

emergence of life on Earth.

2. Challenge our philosophical and religious beliefs about the uniqueness of human life and our role in the universe.

3. Catalyze international cooperation and funding for space exploration and research, potentially driving technological advancements and new discoveries.

The search for extraterrestrial life is an ambitious and exciting endeavor that merges scientific curiosity with our innate desire to understand our place in the universe. As we continue to explore distant planets and moons, develop new technologies, and refine our understanding of the conditions necessary for life, we edge closer to answering the age-old question: Are we alone ?

10 facts about the search for extraterrestrial life:

1. The search for extraterrestrial life, also known as astrobiology, is a scientific field that seeks to understand the conditions and processes that could lead to the emergence of life beyond Earth.

2. The discovery of microbial life in extreme environments on Earth, such as in hot springs or deep-sea hydrothermal vents, has provided evidence that life can thrive in even the harshest conditions.

3. NASA's Kepler mission has discovered thousands of exoplanets, or planets orbiting other stars, many of which are located in the habitable zone, where liquid water could exist on their surfaces.

4. The SETI (Search for Extraterrestrial Intelligence) Institute is a non-profit organization dedicated to the search for extraterrestrial intelligence through the detection of radio signals or other evidence of technological activity.

5. The Fermi Paradox is the apparent contradiction between the high probability of the existence of extraterrestrial civilizations and the lack of evidence for their existence.

6. The discovery of organic molecules on Mars and other planets and

moons in our solar system has provided further evidence that the building blocks of life are common in the universe.

7. The study of extremophiles, or organisms that can survive in extreme conditions, has provided insights into the potential habitats for extraterrestrial life, such as the subsurface oceans of icy moons.

8. The James Webb Space Telescope, set to launch in 2021, will be capable of studying the atmospheres of exoplanets for signs of habitability and even the presence of life.

9. The Drake Equation is a mathematical formula used to estimate the number of technologically advanced civilizations that may exist in the Milky Way galaxy.

10. The search for extraterrestrial life is a complex and interdisciplinary field that involves scientists from a variety of fields, including astronomy, geology, biology, and chemistry.

34

The Nature of Time

Time is an integral part of human experience, dictating the ebb and flow of our lives and shaping our perception of reality. Despite its ubiquity, the true nature of time remains one of the most elusive and debated concepts in science and philosophy. In this article, we will delve into the various theories and interpretations of time, exploring its relationship with space, the role of human perception, and the implications of modern physics in our understanding of this enigmatic phenomenon.

Time in Classical Physics

In classical physics, time was traditionally considered to be an absolute and independent quantity, flowing uniformly throughout the universe. This notion was rooted in the works of Sir Isaac Newton, who posited that time and space were separate entities that provided the backdrop for the events of the universe. In this view, time was regarded as a universal constant, moving forward at a steady and unchanging pace.

Einstein's Revolution: Time and Relativity

Albert Einstein's theory of relativity revolutionized our understanding of time by demonstrating that it is intrinsically linked to space, forming a four-dimensional fabric known as spacetime. According to the theory of relativity, time is not a constant but a relative concept that depends on the observer's speed and gravitational environment. Key insights from relativity theory include:

1. Time dilation: Moving clocks tick slower relative to stationary ones, a phenomenon known as time dilation. This effect has been experimentally confirmed through precise atomic clock measurements.

2. Gravitational time dilation: Time also slows down in stronger gravitational fields. This effect, known as gravitational time dilation, has been observed in experiments comparing atomic clocks at different altitudes.

3. The relativity of simultaneity: Events that appear simultaneous to one observer may occur at different times for another observer in relative motion. This insight challenges the notion of an objective, universal "now."

The Arrow of Time and the Role of Entropy

The arrow of time refers to the apparent unidirectional flow of time, always moving from the past to the future. This directionality is often explained by the concept of entropy, a measure of disorder in a system. According to the second law of thermodynamics, the entropy of a closed system will always increase over time. This irreversible increase in disorder is thought to provide the underlying reason for the one-way flow of time.

Time and Human Perception

The perception of time is a complex and multifaceted phenomenon, influenced by factors such as memory, attention, and emotion. While physical theories of time focus on its objective properties, the study of human perception provides insights into the subjective experience of time. Psychological research has revealed that the perception of time can be distorted, with time seeming to speed up or slow down depending on the individual's state of mind and circumstances.

The nature of time remains a subject of ongoing scientific and philosophical inquiry, with modern physics, cognitive science, and philosophy offering complementary perspectives on this enigmatic concept. From the relative nature of time in Einstein's relativity to the subjective experience of time in the human mind, the exploration of time challenges our understanding of reality and encourages us to reevaluate our place in the cosmos. As we continue to probe the mysteries of time, we deepen our appreciation for the intricate tapestry of existence and the fundamental forces that govern our universe.

10 facts about the nature of time:

1. Time is a fundamental concept in physics and is considered one of the four dimensions of spacetime.

2. The flow of time is often perceived as a constant and irreversible progression from past to present to future, but this perception is not universally accepted.

3. The theory of relativity introduced the concept of time dilation, which suggests that time can appear to pass differently for different observers, depending on their relative motion or gravity.

4. The arrow of time is a concept that describes the directionality of time, with events moving from the past to the future, and is linked to the second law of thermodynamics.

5. The concept of time travel has been explored in science fiction, but is currently not possible according to our current understanding of physics.

6. The study of time perception is a field of psychology that seeks to understand how humans perceive and experience the passage of time.

7. Philosophers have debated the nature of time for centuries, with some suggesting that time is an objective feature of the universe, while others suggest that time is a subjective construct of the human mind.

8. The concept of time is closely linked to causality, which is the relation- ship between events in which one event is seen as the cause of another event.

9. The nature of time is a topic of ongoing research in physics, with theories such as loop quantum gravity and string theory attempting to unify the fundamental forces of nature and explain the nature of time.

10. Despite our understanding of time being incomplete, it remains a crucial aspect of our daily lives and plays a fundamental role in our

perception of the world around us.

35

The Secrets of Black Holes

Black holes, celestial objects with gravitational forces so immense that not even light can escape their pull, have long captivated the imaginations of scientists and the public alike. These cosmic enigmas challenge our understanding of physics, confounding our intuitions and pushing the boundaries of what we know about the universe. In this article, we will delve into the formation and properties of black holes, explore the cutting-edge research aimed at unraveling their mysteries, and discuss their significance for our understanding of the cosmos.

Formation and Properties of Black Holes

Black holes are formed when massive stars exhaust their nuclear fuel and undergo gravitational collapse, compressing their cores to create an object with an incredibly strong gravitational field. There are three main types of black holes, classified according to their mass:

1. Stellar-mass black holes: These black holes form from the collapse of massive stars and typically have masses between 3 and 20 times that of our Sun.
2. Intermediate-mass black holes: These black holes have masses ranging from 100 to 1,000 times that of our Sun, and their formation mechanisms are still debated among scientists.
3. Supermassive black holes: These behemoths are millions to billions

of times more massive than our Sun and are thought to reside at the centers of most galaxies, including our own Milky Way.

Black holes possess several distinctive properties:

1. Event horizon: The boundary around a black hole beyond which nothing, not even light, can escape the gravitational pull. The event horizon marks the point of no return for in-falling matter and energy.
2. Singularity: The infinitely dense core of a black hole, where the laws of physics as we know them break down.
3. Hawking radiation: Theoretical physicist Stephen Hawking predicted that black holes emit radiation due to quantum effects near the event horizon. This process, known as Hawking radiation, causes black holes to lose mass and eventually evaporate over vast timescales.

The Study of Black Holes and Recent Discoveries

The study of black holes has led to groundbreaking discoveries and technological advancements:

1. Gravitational waves: The collision of black holes generates ripples in spacetime known as gravitational waves, which were first detected by the LIGO and Virgo observatories in 2015. This discovery has opened a new window into the universe, allowing scientists to study cosmic events through an entirely different medium.
2. Black hole imaging: In 2019, the Event Horizon Telescope (EHT) collaboration unveiled the first-ever image of a black hole's event horizon, capturing the shadow of the supermassive black hole at the center of galaxy M87. This remarkable achievement provided visual confirmation of black holes' existence and reinforced the predictions of general relativity.
3. Accretion disks and jets: Observations of black holes have revealed that they can be surrounded by rotating disks of in-falling matter known as accretion disks. In some cases, these disks launch relativistic jets of

plasma thousands of light-years into space, providing valuable insights into the complex processes occurring in the vicinity of black holes.

Black Holes and Fundamental Physics

Black holes provide a unique testing ground for theories of fundamental physics, such as general relativity and quantum mechanics. The extreme conditions found near black holes present challenges for our understanding of gravity, spacetime, and the quantum realm. By studying these cosmic enigmas, scientists hope to uncover new insights into the nature of the universe and potentially pave the way for a unified theory that reconciles general relativity with quantum mechanics.

10 facts about the secrets of black holes:

1. Black holes are regions of spacetime where gravity is so strong that nothing, not even light, can escape their grasp.
2. Black holes are formed from the collapse of massive stars, where the core of the star collapses in on itself, creating a singularity at the center.
3. The boundary around a black hole where the gravitational pull is strong enough to trap light is called the event horizon.
4. The study of black holes is crucial for understanding the nature of gravity, as well as the behavior of matter and energy in extreme environments.
5. The concept of Hawking radiation, proposed by physicist Stephen Hawking, suggests that black holes emit radiation and eventually evaporate over time.
6. The study of gravitational waves, or ripples in spacetime caused by the acceleration of massive objects, has provided evidence for the existence of black holes.
7. The gravitational pull of a black hole can cause extreme tidal forces on objects that come too close, stretching them out and tearing them apart in a process known as spaghettification.
8. The study of black holes has led to new discoveries about the structure and evolution of galaxies, as black holes are thought to play a central

role in the formation of galaxies and the growth of supermassive black holes at their centers.

9. The recent discovery of intermediate-mass black holes, which are between 100 and 100,000 times the mass of the sun, has raised new questions about the formation and evolution of black holes.

10. Despite our growing understanding of black holes, many mysteries remain, including the nature of their singularities, the behavior of matter inside the event horizon, and the relationship between black holes and the larger structure of the universe.

36

The Fate of the Universe: Big Crunch or Big Freeze?

The ultimate fate of the universe has been a subject of fascination and debate among scientists and philosophers for centuries. Observations and discoveries in modern cosmology have led to various theories about the possible outcomes of our expanding cosmos. Two widely-discussed scenarios are the Big Crunch, a cataclysmic contraction of the universe, and the Big Freeze, a slow dissipation of energy over time. In this article, we will explore the scientific basis for these theories and discuss the factors that may determine the final destiny of the universe.

Cosmic Expansion and the Role of Dark Energy

The fate of the universe is inextricably linked to its expansion. In 1929, astronomer Edwin Hubble made the groundbreaking discovery that galaxies are receding from each other, indicating that the universe is expanding. Later observations revealed that the expansion is not only ongoing but accelerating. This acceleration is thought to be driven by a mysterious force known as dark energy, which comprises approximately 68% of the universe's total mass-energy content.

The Big Crunch Scenario

The Big Crunch is a hypothetical scenario in which the universe eventually stops expanding and begins to contract, ultimately collapsing back into a hot, dense state akin to the conditions that existed just after the Big Bang. This cosmic contraction would result in a cataclysmic end to the universe, with galaxies, stars, and even atoms being crushed together in a final singularity.

The likelihood of the Big Crunch depends on the density of the universe and the behavior of dark energy. If the universe's density is sufficiently high or if dark energy were to become attractive instead of repulsive, the gravitational forces could overcome the cosmic expansion, leading to a contracting universe.

The Big Freeze Scenario

An alternative fate for the universe is the Big Freeze, also known as the heat death of the universe. This scenario envisions a universe that continues to expand indefinitely, with the rate of expansion increasing due to dark energy's influence. Over time, the universe would become increasingly cold, dilute, and dark as galaxies drift farther apart, stars exhaust their nuclear fuel, and the residual energy dissipates into the vastness of space.

In the Big Freeze scenario, the universe would gradually approach a state of maximum entropy, where all available energy has been evenly distributed and no further work can be done. This ultimate state of thermodynamic equilibrium would mark the end of the universe as a dynamic, evolving system, leading to a cold and lifeless cosmos.

The Balance Between Expansion and Density

The ultimate fate of the universe hinges on the balance between its expansion and density. Observations of the cosmic microwave background radiation, supernovae, and galaxy distributions suggest that the universe's density is currently insufficient to halt its expansion. Under these conditions, the Big Freeze scenario appears to be the more likely outcome.

However, our understanding of the universe is constantly evolving, and future discoveries may yet reveal new insights into the nature of dark energy, the density of the universe, and the fundamental laws that govern its behavior. These advances could potentially shed new light on the fate of the cosmos and refine our understanding of its ultimate destiny.

The question of whether the universe will ultimately succumb to a Big

Crunch or a Big Freeze remains an open one, with the answer dependent on the interplay between cosmic expansion, dark energy, and the universe's density. As scientists continue to probe the depths of the cosmos and unravel the mysteries of its underlying structure, we may one day achieve a more definitive understanding of the universe's ultimate fate. Until then, the cosmic future remains a tantalizing enigma, inspiring us to ponder the grand sweep of cosmic history and the forces that govern the destiny of the cosmos.

10 facts about the fate of the universe:

1. The ultimate fate of the universe is a subject of ongoing scientific inquiry and debate, and there are two main theories: the Big Crunch and the Big Freeze.
2. The Big Crunch theory proposes that the universe will eventually stop expanding and begin to contract, eventually collapsing in on itself in a catastrophic event known as the Big Crunch.
3. The Big Freeze theory, also known as the Heat Death of the Universe, proposes that the universe will continue to expand indefinitely, with galaxies moving further and further apart until they become isolated and dark.
4. The fate of the universe depends on several factors, including the amount of matter and energy in the universe, the expansion rate of the universe, and the nature of dark matter and dark energy.
5. The discovery of dark energy in the late 1990s, which is thought to be responsible for the accelerating expansion of the universe, has provided evidence in favor of the Big Freeze theory.
6. The ultimate fate of the universe may also depend on the geometry of the universe, with a flat universe favoring the Big Freeze theory and a closed universe favoring the Big Crunch theory.
7. The study of cosmic microwave background radiation, which is believed to be leftover radiation from the Big Bang, has provided insights into the early stages of the universe and may provide clues about its ultimate fate.

8. The fate of the universe is a subject of intense study in astrophysics and cosmology, with new discoveries and insights continually emerging.

9. The possibility of a multiverse, or the existence of multiple parallel universes, is a topic of debate among physicists and cosmologists, which could have implications for the fate of our own universe.

10. Regardless of its ultimate fate, the study of the universe and its evolution provides insights into the nature of matter, energy, and the fundamental laws of physics.

37

The Secret of Dark Energy: Fueling the Expansion of the Universe

Dark energy, a mysterious and elusive force that pervades the cosmos, has emerged as one of the most compelling enigmas in modern physics. This enigmatic force is thought to be responsible for the observed acceleration of the universe's expansion, a phenomenon that has challenged our understanding of the cosmos and its ultimate fate. In this article, we will delve into the discovery and properties of dark energy, discuss its role in shaping the universe, and explore ongoing research efforts aimed at unlocking its secrets.

The Discovery of Dark Energy

The existence of dark energy was first inferred in the late 1990s through observations of distant supernovae. These stellar explosions, known as Type Ia supernovae, serve as cosmic mile markers, allowing astronomers to measure the expansion rate of the universe. The discovery that these supernovae were farther away than expected indicated that the universe's expansion was accelerating, contrary to the prevailing view that gravity should cause the expansion to slow down over time.

This unexpected finding prompted the introduction of dark energy as a theoretical force that counteracts gravity, driving the universe's accelerating expansion. Today, dark energy is believed to constitute approximately 68%

of the total mass-energy content of the universe, with dark matter and ordinary matter comprising the remaining 27% and 5%, respectively.

Properties and Theories of Dark Energy

The exact nature of dark energy remains a topic of intense debate and speculation among physicists. Several theories have been proposed to explain its properties and behavior, including:

1. Cosmological constant: Initially proposed by Albert Einstein in the context of general relativity, the cosmological constant represents a form of vacuum energy that permeates all of space. This constant has been reintroduced as a possible explanation for dark energy, with its energy density remaining constant as the universe expands.

2. Quintessence: This theory posits that dark energy is a dynamic, evolving field that changes over time. Quintessence models suggest that the energy density of dark energy may vary across cosmic history, potentially affecting the rate of the universe's expansion in different ways.

3. Modified gravity theories: Some researchers have proposed that the observed acceleration of the universe's expansion may not be due to dark energy, but rather a consequence of modifications to the laws of gravity. These theories explore the possibility that gravity behaves differently on cosmic scales than it does in local environments, potentially accounting for the observed acceleration without invoking dark energy.

Dark Energy and the Fate of the Universe

The presence and properties of dark energy have profound implications for the ultimate fate of the universe. If dark energy remains constant or increases in strength over time, the universe's expansion will continue to accelerate, leading to a future dominated by the Big Freeze scenario. In this outcome, the universe becomes increasingly cold, dilute, and dark as galaxies drift farther apart, stars exhaust their nuclear fuel, and the residual energy dissipates into the vastness of space.

Conversely, if dark energy were to weaken or become attractive over time, the universe's expansion could potentially slow down, halt, or even reverse, leading to a Big Crunch or other scenarios where the universe ultimately contracts.

Unlocking the Secrets of Dark Energy

Numerous experiments and observatories are currently working to decipher the mysteries of dark energy. Projects such as the Dark Energy Survey, the Euclid mission, and the Large Synoptic Survey Telescope aim to map the distribution of galaxies and the cosmic microwave background radiation, providing valuable insights into the universe's expansion history and the properties of dark energy.

Dark energy remains one of the most intriguing and enigmatic aspects of our universe, fueling the expansion of the cosmos and shaping its ultimate destiny.

10 facts about the secret of dark energy:

1. Dark energy is a form of energy that is thought to be responsible for the accelerating expansion of the universe.
2. Dark energy was first proposed in the late 1990s, based on observations of distant supernovae that showed the universe was expanding at an accelerating rate.
3. The nature of dark energy is not well understood, but it is thought to be a property of space itself, rather than a type of matter.
4. Dark energy is estimated to make up approximately 68% of the total energy density of the universe, while ordinary matter makes up only about 5%.
5. The discovery of dark energy has provided evidence for the existence of a cosmological constant, a term in Albert Einstein's theory of general relativity that describes a repulsive force that counteracts gravity.
6. The study of dark energy is crucial for understanding the ultimate fate of the universe, as the accelerating expansion may lead to a "Big Freeze" scenario in which the universe continues to expand and eventually

becomes cold and dark.

7. There are several theories that attempt to explain the nature of dark energy, including the cosmological constant, quintessence, and modified gravity.

8. The study of large-scale structure in the universe, including galaxy clusters and cosmic microwave background radiation, has provided further evidence for the existence of dark energy.

9. The study of dark energy is a subject of ongoing research in astrophysics and cosmology, with new discoveries and insights continually emerging.

10. The nature of dark energy remains one of the greatest mysteries in modern science, and its discovery has challenged our understanding of the fundamental nature of the universe.

38

The Enigma of Cosmic Inflation: How Did the Universe Expand So Quickly?

Cosmic inflation, a key concept in modern cosmology, is a proposed period of rapid expansion that occurred shortly after the Big Bang. This extraordinary expansion is believed to have stretched the universe to a size far larger than we can observe today, addressing several puzzles in the standard cosmological model and setting the stage for the subsequent evolution of the cosmos. In this article, we will explore the origins and implications of cosmic inflation, discuss the observational evidence supporting this theory, and delve into the ongoing research efforts aimed at unlocking the enigma of this early cosmic epoch.

The Origins of Cosmic Inflation

Cosmic inflation was first proposed in the early 1980s by theoretical physicist Alan Guth in response to several problems in the standard Big Bang model. These issues included the flatness problem, the horizon problem, and the magnetic monopole problem. Inflationary theory posits that the universe underwent a brief but dramatic period of exponential expansion within the first fraction of a second after the Big Bang, stretching the fabric of spacetime by a factor of at least 10^{78} in volume.

This rapid expansion is thought to have been driven by a field known as the inflaton field, which possessed a high energy density and exerted a strong repulsive force. As the inflaton field decayed, it released its energy into the universe, reheating it and giving rise to the particles and radiation that constitute the cosmos we observe today.

Observational Evidence for Cosmic Inflation

Several lines of observational evidence support the theory of cosmic inflation:

1. Cosmic microwave background (CMB) radiation: The CMB is the afterglow of the Big Bang, an almost uniform radiation field that fills the universe. Observations of the CMB have revealed subtle fluctuations in temperature, known as anisotropies, that are consistent with the predictions of inflationary theory.

2. Large-scale structure of the universe: Inflation provides a natural explanation for the observed distribution of galaxies and galaxy clusters in the universe, as the tiny fluctuations in the inflaton field during inflation would have seeded the formation of large-scale cosmic structures.

3. Flatness of the universe: The accelerated expansion of the universe during inflation would have flattened the overall curvature of space-time, explaining why the universe appears to be remarkably flat on large scales.

Challenges and Ongoing Research

Despite its successes, cosmic inflation is not without its challenges and open questions. Some of the key areas of ongoing research include:

1. Identifying the inflaton field: The inflaton field is a hypothetical entity that has yet to be conclusively identified or directly observed. Particle physicists and cosmologists are working to develop and test models that could reveal the nature of the inflaton field and its relationship to the known particles and forces.

2. Searching for primordial gravitational waves: Cosmic inflation is predicted to have generated primordial gravitational waves, ripples in

spacetime that carry information about the early universe. Detecting these elusive signals would provide strong evidence for inflation and shed light on the mechanisms that drove the expansion.

3. Exploring alternatives to inflation: While inflation has emerged as the leading explanation for the early universe's rapid expansion, alternative theories continue to be developed and explored. These models, such as cyclic cosmology and bouncing cosmologies, offer alternative perspectives on the early universe and its evolution.

The enigma of cosmic inflation remains at the forefront of modern cosmology, offering a compelling explanation for the early universe's rapid expansion and the observed properties of the cosmos. As researchers continue to probe the depths of the universe and search for clues to its earliest moments, the mysteries of inflation may one day yield to a more comprehensive understanding of the origins and evolution of the cosmos.

10 facts about the enigma of cosmic inflation:

1. Cosmic inflation is a theory that proposes that the universe underwent a brief period of exponential expansion immediately after the Big Bang.

2. The theory of cosmic inflation was first proposed in the 1980s to explain several mysteries of the early universe, including the uniformity of the cosmic microwave background radiation and the absence of magnetic monopoles.

3. Cosmic inflation is thought to have occurred approximately 10^{-36} seconds after the Big Bang and lasted for a fraction of a second.

4. During cosmic inflation, the universe is thought to have expanded by a factor of at least 10^{26}, from an infinitesimal size to approximately the size of a grapefruit.

5. The mechanism behind cosmic inflation is still not well understood, but it is thought to have been caused by a scalar field, or inflaton, that permeated the early universe.

6. The study of the cosmic microwave background radiation, which is believed to be leftover radiation from the Big Bang, has provided strong evidence in support of the theory of cosmic inflation.

7. The concept of eternal inflation, which proposes the existence of an infinite number of universes that are constantly being created, has been linked to the theory of cosmic inflation.

8. The study of cosmic inflation is a subject of ongoing research in astrophysics and cosmology, with new discoveries and insights continually emerging.

9. Cosmic inflation is thought to have played a crucial role in the formation of the large-scale structure of the universe, including the formation of galaxies and galaxy clusters.

10. Despite its potential importance in our understanding of the early universe, the mechanism behind cosmic inflation remains one of the greatest unsolved mysteries in modern physics.

39

The Power of Gravitational Waves: Listening to the Universe

Gravitational waves, ripples in the fabric of spacetime caused by the acceleration of massive objects, have emerged as a groundbreaking tool in the field of astronomy. These elusive waves, first predicted by Albert Einstein in 1916 as a consequence of his general theory of relativity, provide a unique window into the universe, allowing scientists to study phenomena that would otherwise be inaccessible through traditional observational methods. In this article, we will explore the discovery and detection of gravitational waves, discuss their significance for our understanding of the cosmos, and delve into the future prospects for this exciting area of research.

The Discovery of Gravitational Waves

The existence of gravitational waves remained purely theoretical for nearly a century after Einstein's prediction, as their detection posed immense technical challenges. However, in September 2015, the Laser Interferometer Gravitational-Wave Observatory (LIGO) made a groundbreaking discovery: the direct observation of gravitational waves for the first time. This historic detection, announced in February 2016, was the result of the merger of two black holes more than a billion light-years away.

Since then, LIGO, along with its European counterpart Virgo, has detected numerous additional gravitational wave events, including the mergers of neutron stars and black holes. These observations have ushered in a new era of astronomy known as gravitational-wave astronomy.

The Significance of Gravitational Waves

Gravitational waves carry invaluable information about the universe and its constituents, allowing scientists to probe phenomena that are otherwise difficult or impossible to study using electromagnetic radiation. Some of the key insights enabled by gravitational-wave observations include:

1. Testing general relativity: The detection of gravitational waves provides a powerful test of Einstein's general theory of relativity in the strong-field regime, where the gravitational forces are most intense. To date, these observations have been consistent with the predictions of general relativity, confirming its validity as a description of gravity.

2. Probing the properties of black holes and neutron stars: The mergers of black holes and neutron stars produce strong gravitational wave signals, enabling researchers to study these enigmatic objects in detail. Gravitational-wave observations have provided insights into the masses, spins, and population statistics of black holes and neutron stars, shedding light on their formation and evolution.

3. Investigating the early universe: Gravitational waves from the early universe, if detected, could offer a unique glimpse into the conditions that prevailed during the first moments after the Big Bang. In particular, the observation of primordial gravitational waves would provide strong evidence for the theory of cosmic inflation.

The Future of Gravitational-Wave Astronomy

The field of gravitational-wave astronomy is still in its infancy, with numerous ongoing and planned projects aimed at expanding our understanding of the universe through the detection and analysis of gravitational waves:

1. Improved ground-based detectors: Upgrades to existing facilities, such as LIGO and Virgo, as well as the construction of new detectors, such as LIGO-India and the Kamioka Gravitational Wave Detector

(KAGRA) in Japan, will enhance the sensitivity and reach of ground-based gravitational-wave observatories.

2. Space-based observatories: The planned Laser Interferometer Space Antenna (LISA) mission, set for launch in the 2030s, will be sensitive to lower-frequency gravitational waves than ground-based detectors, enabling the study of a wider range of astrophysical phenomena, including the mergers of supermassive black holes and the formation of compact binary systems.

Pulsar timing arrays: By monitoring the precise timing of radio signals from an array of distant pulsars, researchers aim to detect low-frequency gravitational waves produced by the mergers of supermassive black holes in distant galaxies Other applications: Gravitational-wave astronomy has already led to new technologies and innovations, including advances in data analysis, computing, and precision measurement, with potential applications in fields ranging from medicine to national security.

Multi-messenger astronomy: The study of gravitational waves in conjunction with other forms of radiation, such as light and neutrinos, known as multi-messenger astronomy, has the potential to provide unprecedented insights into the behavior of the universe and its fundamental laws.

Cosmic origins explorer (CORE): The proposed space-based mission, CORE, aims to detect primordial gravitational waves generated in the first fractions of a second after the Big Bang, providing insights into the nature of the universe in its earliest stages.

The future of gravitational-wave astronomy is bright, with new discoveries and insights expected to emerge in the coming decades as new observatories and technologies come online and more data is analyzed.

10 facts about the power of gravitational waves:

1. Gravitational waves are ripples in the fabric of spacetime caused by the

acceleration of massive objects, such as the collision of two black holes or neutron stars.

2. Gravitational waves were first predicted by Albert Einstein's theory of general relativity in 1915, but were not directly detected until 2015.

3. The detection of gravitational waves was made possible by the Laser Interferometer Gravitational-Wave Observatory (LIGO), a collaborative project between MIT and Caltech.

4. Gravitational waves are incredibly difficult to detect, as they cause only very tiny distortions in spacetime, but the detection of these waves has opened up a new window into the universe.

5. Gravitational waves can be used to study a wide range of astrophysical phenomena, including the behavior of black holes and neutron stars, the formation of galaxies, and the nature of dark matter and dark energy.

6. The study of gravitational waves has already led to several major discoveries, including the first direct detection of a binary black hole merger and the first detection of a binary neutron star merger.

7. Gravitational wave astronomy is a rapidly growing field, with new detectors and observing techniques being developed to further our understanding of the universe.

8. The study of gravitational waves has the potential to revolutionize our understanding of the universe, and may even provide insights into the nature of quantum gravity and the origins of the universe.

9. The detection of gravitational waves has also led to new technologies and innovations, such as advanced computing and data analysis techniques.

10. The study of gravitational waves is a collaborative effort between scientists and institutions from around the world, and represents one of the most exciting frontiers in modern astrophysics.

40

The Secret of Neutron Stars: The Densest Objects in the Universe

Neutron stars, the remnants of massive stars that have undergone a supernova explosion, are among the most fascinating and extreme objects in the universe. These ultra-dense, city-sized objects pack a mass up to twice that of our Sun into a radius of just about 10 kilometers. In this article, we will explore the formation and properties of neutron stars, discuss their role in the cosmos, and delve into the ongoing research efforts aimed at unraveling the mysteries of these extraordinary celestial bodies.

The Formation of Neutron Stars

Neutron stars are born in the aftermath of a supernova, a powerful explosion that marks the end of the life cycle of massive stars. When a star with a mass between about 1.4 and 3 times that of the Sun exhausts its nuclear fuel, it can no longer support itself against the crushing force of gravity. As a result, the core of the star collapses in a fraction of a second, compressing protons and electrons into neutrons and forming a dense core primarily composed of neutrons.

The outer layers of the star are ejected in the supernova explosion, leaving behind the newly-formed neutron star. If the progenitor star's mass is above approximately 3 solar masses, the gravitational forces at play are so intense that the core collapses further, ultimately forming a black hole.

Properties of Neutron Stars

Neutron stars exhibit a remarkable array of properties that make them unique objects of study:

1. Density: Neutron stars are incredibly dense, with a teaspoon of neutron star material weighing as much as a mountain, around a billion tons. The density of a neutron star is so extreme that it is comparable to the density of atomic nuclei.

2. Magnetic fields: Neutron stars possess immensely strong magnetic fields, often billions of times stronger than the Earth's magnetic field. Some neutron stars, known as magnetars, have magnetic fields that are even stronger, reaching up to a thousand trillion times that of Earth's.

3. Rotation: Neutron stars spin rapidly, with some rotating hundreds of times per second. This rapid rotation, combined with their strong magnetic fields, can cause neutron stars to emit intense beams of electromagnetic radiation from their magnetic poles. When these beams sweep across Earth, they produce a periodic signal known as a pulsar.

The Role of Neutron Stars in the Cosmos

Neutron stars play a crucial role in various astrophysical processes and phenomena:

1. Nucleosynthesis: The powerful supernova explosions that give birth to neutron stars are responsible for creating many of the heavier elements found in the universe, including elements such as gold, platinum, and uranium.

2. Gravitational waves: The merger of two neutron stars can generate detectable gravitational waves, ripples in spacetime that provide valuable insights into the properties of these extreme objects and the nature of gravity itself.

3. Cosmic laboratories: Neutron stars serve as cosmic laboratories for studying the behavior of matter under extreme conditions, providing

insights into the fundamental forces and particles that govern the universe.

Unraveling the Mysteries of Neutron Stars

Ongoing research efforts are focused on furthering our understanding of neutron stars and their properties:

1. Observational studies: Observations of neutron stars across the electro-magnetic spectrum, from radio waves to gamma rays, provide valuable information about their magnetic fields, rotation, and emission pro-cesses.
2. Gravitational-wave astronomy: The study of gravitational waves from neutron star mergers offers a unique opportunity to probe the interior structure and properties of these enigmatic objects.

3.Theoretical modeling: Researchers are developing increasingly sophisti-cated models of neutron stars, incorporating the latest advances in nuclear physics, particle physics, and general relativity to better understand the behavior of matter under such extreme conditions.

4.Multi-messenger astronomy: Combining information from gravita-tional waves, electromagnetic radiation, and neutrinos, multi-messenger astronomy allows for a comprehensive understanding of neutron star mergers and their astrophysical implications.

5.Equation of state: One of the most significant open questions in neutron star research is the determination of the equation of state, which describes the relationship between pressure, density, and temperature in neutron star matter. The equation of state is crucial for understanding the internal structure of neutron stars and predicting their maximum mass and radius.

Neutron stars, the densest objects in the universe, provide a unique window into the extreme conditions that prevail in the cosmos. As researchers continue to observe, model, and study these enigmatic celestial bodies, they will undoubtedly unlock new secrets and deepen our understanding of the fundamental forces and particles that shape the universe. With the ongoing advancements in observational and theoretical techniques, the secrets of neutron stars will continue to be unraveled, shedding light on the nature of matter and the fabric of spacetime itself.

10 facts about the secret of neutron stars:

1. Neutron stars are the collapsed cores of massive stars that have undergone a supernova explosion, with a mass approximately 1.4 times that of the sun, but a radius of only about 10 km.
2. Neutron stars are incredibly dense, with a mass equivalent to that of the entire sun but a volume that is about a billion times smaller.
3. Neutron stars are composed primarily of neutrons, with a thin outer layer of iron, nickel, and other heavy elements.
4. The extreme density of neutron stars results in a surface gravity that is about 2 billion times stronger than that on Earth.
5. Neutron stars are believed to have strong magnetic fields, which can result in the emission of radiation, including X-rays and gamma rays.
6. The study of pulsars, which are rapidly rotating neutron stars that emit beams of radiation, has provided insights into the properties and behavior of neutron stars.
7. The study of neutron stars has led to new discoveries and insights into fundamental physics, including the behavior of matter at extreme densities and the nature of the strong force.
8. The discovery of gravitational waves from the merger of two neutron stars in 2017 provided evidence for the origin of heavy elements, such as gold and platinum, in the universe.
9. Neutron stars are thought to play a crucial role in the formation of binary systems, where two stars orbit around a common center of mass.
10. The study of neutron stars is a subject of ongoing research in astrophysics and cosmology, with new discoveries and insights continually emerging.

V

Secrets of History

41

The True Identity of Jack the Ripper

Jack the Ripper is one of the most infamous serial killers in history, known for his brutal and gruesome murders in the Whitechapel district of London in the late 19th century. Despite numerous investigations and suspects over the years, the true identity of Jack the Ripper remains a mystery. In this article, we will explore the various theories and suspects surrounding Jack the Ripper and examine the evidence that has been gathered over the years.

The Murders

Between August and November 1888, Jack the Ripper is believed to have killed at least five women in the Whitechapel district of London. The victims were all prostitutes, and their bodies were mutilated and left in public areas. The brutality of the murders, along with the fact that the killer was never caught, has captivated people's imaginations for over a century.

The Suspects

Over the years, many suspects have been identified as potential Jack the Ripper candidates, but none have been definitively proven to be the killer. Some of the most famous suspects include:

1. Aaron Kosminski: A Polish immigrant who lived in Whitechapel at the time of the murders, Kosminski was identified as a suspect by police at the time. However, there is little evidence to support his guilt, and he was never formally charged.

2. Sir William Gull: A prominent physician who was suspected by some as being involved in the murders. However, there is little evidence to support this theory, and it is considered unlikely.

3. Montague John Druitt: A barrister who was found dead shortly after the last murder, and who some have suggested may have been the killer. However, there is no solid evidence to support this theory.

4. Francis Tumblety: A physician and known misogynist who was in London at the time of the murders. He was briefly arrested but released due to lack of evidence.

The Evidence

Despite numerous investigations and suspects over the years, the true identity of Jack the Ripper remains a mystery. The lack of concrete evidence and the fact that the murders took place over 130 years ago have made it difficult to definitively identify the killer. Some of the evidence that has been gathered over the years includes:

1. Letters: The killer sent several letters to the police and media, including the famous "From Hell" letter. However, the authenticity of these letters is disputed, and it is unclear whether they were actually written by the killer.

2. Crime Scenes: The police gathered evidence at the crime scenes, including bloodstains and fingerprints. However, forensic technology at the time was limited, and much of the evidence has since been lost or destroyed.

3. Witness Statements: Several witnesses came forward claiming to have seen the killer or the victims on the night of the murders. However, witness testimony can be unreliable, and much of the information provided by witnesses conflicted with other accounts.

Jack the Ripper remains one of the most enigmatic and mysterious figures in history. Despite numerous investigations and suspects over the years, the true identity of the killer remains a mystery. While new theories and evidence may emerge in the future, it is likely that the true identity of Jack the Ripper will remain a mystery, continuing to fascinate and intrigue people for generations to come.

42

Prophet Moses (Musa) and the Ark of the Covenant

The Ark of the Covenant, a sacred relic described in the Hebrew Bible, is an enduring mystery that has captivated historians, archaeologists, and religious scholars for centuries. Associated with the biblical prophet Moses (Musa in Islam), the Ark is said to contain the stone tablets inscribed with the Ten Commandments, along with other sacred items. Despite numerous attempts to locate this mysterious artifact, the Ark's current whereabouts remain unknown. In this article, we will explore the historical and religious significance of the Ark of the Covenant, various theories about its location, and the ongoing search for this enigmatic relic.

Historical and Religious Significance

The Ark of the Covenant is an important artifact in the Abrahamic religions, particularly Judaism and Christianity. According to the Hebrew Bible, the Ark was constructed at the command of God, who provided detailed instructions to Moses on its design and purpose. The Ark served as a symbol of God's presence and protection, as well as a means of communicating with Him. It played a central role in the religious and military exploits of the ancient Israelites and was eventually placed in the Holy of Holies, the innermost chamber of Solomon's Temple in Jerusalem.

Various Theories About the Ark's Location

Throughout history, numerous theories have emerged regarding the location of the Ark of the Covenant. Some of the most prominent theories include:

1. Hidden in Jerusalem: Some scholars believe that the Ark was hidden beneath the Temple Mount in Jerusalem before the Babylonian conquest in 587 BCE. This theory suggests that the Ark remains concealed in a secret chamber or tunnel system, awaiting discovery.

2. Taken to Ethiopia: According to Ethiopian tradition, the Ark was brought to the city of Axum by Menelik I, the legendary first ruler of Ethiopia and son of King Solomon and the Queen of Sheba. The Ark is said to be housed in the Church of Our Lady Mary of Zion in Axum, where it is guarded by a select group of priests who are forbidden to reveal it to the public.

3. Hidden in Southern Africa: Another theory posits that the Ark was taken to southern Africa by a group of Israelites who migrated there after the destruction of Solomon's Temple. Proponents of this theory point to the discovery of the Lemba people, a Bantu-speaking group with Semitic ancestry who claim to be descendants of ancient Israelites and possess a sacred object resembling the Ark.

4. Lost or destroyed: Some scholars argue that the Ark was either lost or destroyed during the tumultuous history of the ancient Israelites, particularly during the Babylonian conquest or the Roman destruction of the Second Temple in 70 CE.

The Ongoing Search for the Ark

Despite numerous expeditions and investigations, the Ark of the Covenant remains one of history's most elusive artifacts. Archaeologists, historians, and religious scholars continue to search for clues that might shed light on the Ark's location or fate. The quest for the Ark is driven by a combination of religious, historical, and cultural motivations, as the discovery of the Ark would have significant implications for our understanding of the ancient world and the Abrahamic religions.

The mystery of the Ark of the Covenant of Prophet Moses (Musa) remains unsolved, with various theories suggesting different locations and fates for the sacred artifact. As the search continues, the Ark remains a powerful symbol of faith, a testament to the enduring power of religious and historical narratives. Whether the Ark will ever be found is uncertain, but its enigmatic allure ensures that it will continue to captivate the imaginations of scholars and believers alike.

43

The Curse of Tutankhamun's Tresure

The discovery of Tutankhamun's tomb in 1922 by British archaeologist Howard Carter sparked a worldwide fascination with the young pharaoh and his treasures. Along with this fascination, a series of mysterious deaths and misfortunes associated with those who entered the tomb gave rise to the legend of the Curse of Tutankhamun. In this article, we will explore the history of the discovery, the alleged curse, and the scientific explanations for the strange events that surrounded the unearthing of the ancient Egyptian treasure.

The Discovery of Tutankhamun's Tomb

Tutankhamun, who reigned during the 18th dynasty of the New Kingdom (circa 1334-1325 BCE), was a relatively minor pharaoh whose name and reign were largely forgotten by history. However, the discovery of his intact tomb in the Valley of the Kings by Howard Carter and his patron, Lord Carnarvon, catapulted the young pharaoh to global fame.

Tutankhamun's tomb was exceptional, as it had remained undisturbed by tomb robbers and was filled with priceless artifacts, including the iconic gold funerary mask. The treasures and well-preserved nature of the tomb provided archaeologists with a unique glimpse into the burial practices and material wealth of the ancient Egyptians.

The Curse of Tutankhamun

The legend of the curse began shortly after the tomb's discovery when Lord

Carnarvon died under mysterious circumstances in Cairo, just a few months after the opening of the tomb. Rumors spread that he had succumbed to the curse, fueled by reports of an inscription found in the tomb that read: "Death shall come on swift wings to him who disturbs the peace of the king."

Following Carnarvon's death, a series of misfortunes and deaths occurred among those who were involved in the excavation or had visited the tomb, further perpetuating the myth of the curse. Notable incidents include the deaths of several members of Carter's team, as well as the suicide of Lord Carnarvon's half-brother shortly after his visit to the tomb.

The media played a significant role in popularizing the notion of the curse, with sensational headlines and stories captivating the public's imagination. The legend of the curse has endured to this day, inspiring countless books, movies, and documentaries.

Scientific Explanations for the Curse

In recent years, researchers have proposed several scientific explanations for the mysterious events surrounding the discovery of Tutankhamun's tomb:

1. Bacteria and mold: Some experts suggest that the tomb's sealed environment allowed the growth of potentially harmful bacteria and mold. Exposure to these pathogens could have led to severe respiratory infections, which could explain some of the deaths and illnesses associated with the tomb.

2. Toxic substances: Another theory posits that the tomb's artifacts and wall paintings were coated with toxic substances, such as arsenic or lead-based pigments, which could have caused poisoning in those who came into contact with them.

3. Psychological factors: The power of suggestion and the belief in the curse could have led to a heightened state of anxiety and stress among those involved in the excavation, contributing to accidents, illnesses, and even death.

Conclusion

The Curse of Tutankhamun's Treasure, while a captivating and enduring legend, is likely the result of a combination of natural causes, coincidence, and the power of suggestion. The fascination with ancient Egypt and its enigmatic pharaohs, along with the mysterious circumstances surrounding the tomb's discovery, have ensured that the myth of the curse continues to captivate the public's imagination. However, scientific explanations

44

The Mystery of the Prophet Sulaiman Temple

The Temple of the Prophet Sulaiman, also known as Solomon's Temple, is a subject of intrigue and fascination for historians, archaeologists, and religious scholars alike. The temple, which is believed to have been built by the biblical King Solomon, son of King David, has captivated the imagination of countless generations due to its association with legendary stories and its enigmatic history. In this article, we will explore the religious and historical background of the temple, discuss the various theories surrounding its location and construction, and delve into the ongoing quest to unravel the mysteries of this ancient site.

Religious and Historical Background

The Temple of the Prophet Sulaiman is mentioned in various religious texts, including the Hebrew Bible, the Christian Old Testament, and the Islamic Quran. According to these sources, King Solomon constructed the temple as a place of worship and as a repository for the Ark of the Covenant, which contained the original stone tablets inscribed with the Ten Commandments. The temple was purportedly an architectural marvel, adorned with gold, silver, and precious stones, and served as the spiritual center of the Israelites.

The temple's construction is believed to have taken place around the 10th century BCE. However, it was destroyed by the Babylonians in 586 BCE during their conquest of Jerusalem. Subsequent attempts were made to rebuild the temple, most notably by the Persians and later by the Romans, but none of these structures endured the test of time.

Theories Surrounding the Temple's Location and Construction

Despite the temple's prominence in religious texts, its exact location remains a subject of debate among scholars. Some theories suggest that it was built on the Temple Mount in Jerusalem, which is currently home to the Islamic holy sites of Al-Aqsa Mosque and the Dome of the Rock. Others believe that the temple's remains lie hidden beneath these structures or elsewhere in the city. However, definitive archaeological evidence of the temple's existence and location has yet to be discovered, in part due to the political and religious sensitivities surrounding excavations in Jerusalem.

The construction of the temple itself is also shrouded in mystery. According to legend, Solomon enlisted the help of supernatural beings, such as jinn, to aid in the construction of the temple. These beings were said to possess incredible strength and the ability to shape and transport massive stones. While this account is likely rooted in folklore, it has fueled speculation about the advanced techniques that may have been employed in the temple's construction.

The Quest to Unravel the Temple's Mysteries

The search for the Temple of the Prophet Sulaiman and the secrets it may hold continues to this day. Several factors contribute to the ongoing fascination with this enigmatic site:

1. The temple's religious significance: As a central feature in the Abrahamic faiths, the temple holds immense spiritual importance for Jews, Christians, and Muslims alike.
2. The potential archaeological discoveries: Unearthing the temple's remains would be a monumental archaeological achievement, providing invaluable insights into the religious practices, artistic styles, and architectural techniques of the ancient Israelites.
3. The legendary artifacts: The temple is said to have housed numerous

sacred relics, such as the Ark of the Covenant and the Menorah. The discovery of these artifacts would have profound implications for our understanding of the temple's history

The mystery of the Prophet Sulaiman Temple remains an enduring enigma, captivating the imagination of scholars and the public alike.

45

The Secrets of the Knights Templar

The Knights Templar were a medieval Christian military order that played a significant role in European history. Founded in the early 12th century, the order was initially formed to protect Christian pilgrims traveling to the Holy Land, but later became involved in banking and finance, and gained significant political and economic power. Despite their prominence, the Knights Templar were eventually disbanded, and their legacy remains shrouded in mystery and intrigue.

One of the most enduring mysteries surrounding the Knights Templar is their rumored possession of the Holy Grail, a vessel said to have been used by Jesus Christ at the Last Supper. According to legend, the Knights Templar were tasked with protecting the Holy Grail and other relics, including the Ark of the Covenant and the Spear of Destiny. Some believe that the order's wealth and power were tied to these legendary artifacts, and that the Knights Templar may have been involved in secret societies and occult practices.

Another area of intrigue surrounding the Knights Templar is their association with the Freemasons, a fraternal organization that traces its origins back to the 16th century. Some believe that the Knights Templar were instrumental in the formation of the Freemasons, and that the two organizations share a common set of beliefs and rituals. There is also speculation that the Knights Templar may have possessed secret knowledge or technology, such as the ability to harness the power of electricity, that has been lost to history.

The downfall of the Knights Templar is another area of mystery and controversy. In the early 14th century, the order was disbanded by Pope Clement V, and its members were persecuted and accused of heresy. Some believe that the persecution of the Knights Templar was motivated by their wealth and power, while others believe that the order's secrets and esoteric practices made them a threat to the established church and monarchy.

Despite the many legends and rumors surrounding the Knights Templar, the true nature of the order remains elusive. The Knights Templar continue to fascinate historians and conspiracy theorists alike, with new theories and discoveries emerging regularly. The legacy of the Knights Templar, both as a symbol of medieval chivalry and as a source of mystery and intrigue, continues to captivate the imagination of people around the world.

46

The Secret Life of Leonardo da Vinci: The Man Behind the Art

Leonardo da Vinci is widely considered one of the greatest artists and thinkers of the Renaissance era, with works such as the Mona Lisa and The Last Supper capturing the imagination of people around the world. However, the man behind the art was a complex and multifaceted figure, with a life filled with secrets, passions, and contradictions.

One of the most fascinating aspects of Leonardo's life was his status as a homosexual. While homosexuality was considered taboo during his time, Leonardo's writings and personal relationships suggest that he was attracted to both men and women. Some of his most intimate relationships were with his male apprentices and assistants, with whom he shared both intellectual and physical connections.

Leonardo's fascination with science and technology was another key aspect of his life. He was a skilled inventor, designing machines and devices that were far ahead of their time. He also studied anatomy, optics, and geology, making numerous breakthroughs and discoveries that helped pave the way for modern science.

In addition to his scientific pursuits, Leonardo was deeply interested in spirituality and the mysteries of the universe. He was a student of esoteric and occult knowledge, studying the writings of philosophers and mystics from around the world. Some believe that his interest in these subjects may have influenced his art, giving it a deeper spiritual and symbolic meaning.

Leonardo's personal life was also marked by a number of challenges and struggles. He suffered from depression and anxiety throughout his life, and struggled to find acceptance and validation for his work. He also faced financial difficulties, often working on multiple projects simultaneously to support himself and his assistants.

Despite these challenges, Leonardo's legacy as an artist and thinker remains unchallenged. His works continue to inspire and captivate people around the world, with their beauty, depth, and complexity. His life, with its secrets and contradictions, serves as a reminder of the complexity of the human experience, and the power of art and creativity to transcend time and space.

47

The Enigma of the Antikythera Mechanism: A 2000-Year-Old Computer

The Antikythera Mechanism is a 2000-year-old Greek device that has fascinated scholars and scientists since its discovery in 1901. The device is believed to be the world's oldest known analog computer, and is considered a marvel of ancient engineering and technology. Despite decades of study and analysis, the purpose and function of the Antikythera Mechanism remains an enigma.

The device was discovered by a group of sponge divers off the coast of the Greek island of Antikythera in 1901. The divers recovered a number of ancient artifacts from a sunken Roman cargo ship, including the remains of a complex bronze mechanism. The device was initially believed to be a navigational instrument, but subsequent study revealed that it was much more complex than originally thought.

The Antikythera Mechanism consists of a series of interlocking gears and wheels, contained within a wooden and bronze case. The device is believed to have been used to calculate the positions of the sun, moon, and planets, and to predict eclipses and astronomical events. The mechanism was operated by turning a hand-crank, which caused the gears and wheels to move and display astronomical information on a series of dials and pointers.

Despite its age, the Antikythera Mechanism is a remarkable feat of engineering and technology. The device is believed to have been accurate to within a single degree, and to have been able to predict astronomical events with great precision. Its complexity and sophistication have led some to speculate that it may have been built by a team of skilled artisans, working under the guidance of an expert astronomer or mathematician.

The study of the Antikythera Mechanism has provided insights into the history of science and technology, and has challenged our understanding of ancient engineering and astronomy. The device remains a subject of ongoing research and study, with new discoveries and insights emerging regularly. Despite centuries of technological progress, the Antikythera Mechanism stands as a testament to the ingenuity and innovation of our ancient ancestors, and a reminder of the mysteries and wonders of the universe.

48

The Mystery of the Mary Celeste: Ghost Ship or Insurance Fraud?

The Mary Celeste was an American merchant ship that was discovered abandoned in the Atlantic Ocean in 1872. The ship was found adrift, with no crew on board, and with no obvious signs of distress or foul play. The discovery of the Mary Celeste has since become one of the most enduring mysteries of maritime history, with theories ranging from ghost ships to insurance fraud.

The Mary Celeste was originally launched in 1861 as the Amazon, and was later sold and re-named by a succession of owners. The ship was bound for Genoa, Italy, with a cargo of alcohol when it was discovered adrift in the Atlantic Ocean by a passing British ship, the Dei Gratia. The ship's cargo and supplies were largely intact, and there was no sign of struggle or violence on board.

Despite extensive investigations and inquiries, no satisfactory explanation for the disappearance of the Mary Celeste's crew has ever been found. Numerous theories have been proposed, including piracy, mutiny, and foul play, but none of these theories have been proven. Some have suggested that the crew may have been abducted by extraterrestrial beings, while others believe that the ship may have been cursed or haunted.

One of the most widely accepted theories is that the Mary Celeste was the victim of insurance fraud. It has been suggested that the ship's owner, Benjamin Briggs, may have deliberately abandoned the ship in order to collect on the insurance policy. This theory is supported by the fact that Briggs was heavily in debt at the time of the ship's disappearance, and that the ship's cargo was highly valuable.

Despite the popularity of the insurance fraud theory, there is little concrete evidence to support it. The disappearance of the Mary Celeste remains a subject of ongoing debate and speculation, with new theories and discoveries emerging regularly. The mystery of the Mary Celeste continues to captivate the imagination of people around the world, and serves as a reminder of the enduring power of maritime legends and folklore.

49

The Power of the Knights Hospitaller: Protectors of Christianity and the Holy Land

The Knights Hospitaller, also known as the Knights of Malta, were a medieval Christian military order that played a significant role in the defense of Christianity and the Holy Land. Founded in the early 12th century, the order was initially established to provide medical care to Christian pilgrims traveling to the Holy Land, but later became a major military force, fighting against Muslim armies and defending Christian territories.

The Knights Hospitaller were renowned for their skill and bravery in battle, as well as their commitment to their Christian faith. They were known for their distinctive black and white cloaks, emblazoned with the symbol of the order, a white cross on a black background. The order was also notable for its extensive network of hospitals, clinics, and infirmaries, which provided medical care to people throughout Europe and the Holy Land.

During the Crusades, the Knights Hospitaller played a crucial role in the defense of Christian territories in the Holy Land. They fought in numerous battles and sieges, including the Siege of Acre and the Battle of Montgisard, and were instrumental in the recapture of Jerusalem in 1099. The order also established a number of fortified cities and castles in the Holy Land, including Rhodes, which served as the headquarters of the order for over 200 years.

In addition to their military and medical activities, the Knights Hospitaller were also known for their contributions to art, architecture, and culture. They commissioned numerous works of art, including paintings, sculptures, and tapestries, and were patrons of many of the great artists of the Renaissance era.

The Knights Hospitaller eventually declined in power and influence, as Christian forces were gradually pushed out of the Holy Land. The order relocated to various locations throughout Europe, and eventually settled in Malta, where they continued to play a role in European politics and diplomacy. The order was eventually disbanded in the late 18th century, but its legacy as a defender of Christianity and the Holy Land continues to inspire and captivate people around the world.

Today, the Knights Hospitaller are remembered for their bravery, skill, and commitment to their Christian faith. Their contributions to medicine, art, and culture continue to be celebrated, and their legacy as protectors of Christianity and the Holy Land remains a source of inspiration and pride for many.

50

The Secret of the Voynich Manuscript: A Code Yet to Be Broken.

The Voynich Manuscript is one of the most enigmatic and mysterious documents in history. Written in an unknown language, using an unknown script, and filled with bizarre illustrations of plants, animals, and astronomical diagrams, the manuscript has confounded scholars, code-breakers, and amateur sleuths for centuries.

The manuscript is believed to have been created in the early 15th century, although its origins and purpose remain unclear. The document was rediscovered in the early 20th century by a Polish book dealer named Wilfrid Voynich, who acquired it from a Jesuit college in Italy. Since then, the manuscript has been the subject of intense study and speculation, with numerous theories proposed about its origin, content, and meaning.

Despite decades of effort, no one has been able to decipher the text of the Voynich Manuscript. The script is believed to be a type of cipher, possibly based on a complex system of symbols and codes. Some have suggested that the manuscript may be a work of medieval cryptography, designed to conceal secret knowledge or information from prying eyes.

The illustrations and diagrams in the manuscript have also been the subject of intense study and analysis. The drawings depict a variety of plants, animals, and astronomical phenomena, many of which are unknown or poorly understood. Some have suggested that the illustrations may be coded messages, or that they may represent some kind of symbolic language.

The Voynich Manuscript remains a source of fascination and intrigue for scholars and laypeople alike. Despite numerous attempts to decode the manuscript, its secrets remain hidden, and its true purpose and meaning continue to elude us. The manuscript is a reminder of the enduring power of mystery and the unknown, and of the seemingly infinite depths of human knowledge and imagination.

VI

Secrets of Nature

51

The Secret Life of Plants

The idea that plants may have a secret life, beyond what we can see with our eyes, has captured the imagination of people for centuries. While some of the claims made about plants and their supposed abilities may be exaggerated, there is growing evidence that plants are far more complex and sophisticated than we once thought.

One of the most intriguing aspects of plant life is their ability to communicate with one another. Plants have been shown to release chemicals, such as pheromones and other volatile compounds, that can signal to other plants in their vicinity. This chemical signaling allows plants to coordinate their growth and development, and to defend against predators and other threats.

Plants have also been shown to respond to their environment in a variety of ways. They can detect changes in temperature, humidity, and light, and can adjust their growth and behavior accordingly. Some plants have even been shown to respond to music and other forms of sound, with certain frequencies and rhythms appearing to have a positive effect on plant growth and health.

Another area of plant life that has fascinated scientists and laypeople alike is the concept of plant consciousness. While there is no clear consensus on what consciousness means, some scientists and philosophers have suggested that plants may possess some form of consciousness or awareness. This idea is supported by recent research showing that plants have a complex network of sensors and signaling pathways, and that they can adapt their behavior in response to changing conditions.

The potential uses of plant life for medicine and other applications are also a source of fascination. Many plants have been used for centuries in traditional medicine, and recent research has shown that some plants contain powerful compounds that may have therapeutic properties. For example, the compound salicylic acid, found in willow bark, was used as the basis for the development of aspirin.

The secret life of plants continues to be a subject of ongoing research and speculation. While we may never fully understand the inner workings of plant life, the growing body of evidence suggests that plants are far more complex and sophisticated than we once thought, and that they may hold many secrets yet to be discovered.

It is important to note that while some studies have shown that plants can respond positively to certain frequencies and rhythms, there is no definitive list of specific frequencies for individual plants. Additionally, different types of plants may respond differently to the same frequency.

That being said, here is a general list of frequencies from highest to lowest that have been associated with positive effects on plant growth and health:

1. 528 Hz – This frequency is often referred to as the "love frequency" and has been shown to promote plant growth and stress resistance.
2. 432 Hz – This frequency is often referred to as the "natural frequency" and is believed to have a calming effect on plants, promoting overall health and well-being.
3. 639 Hz – This frequency is believed to promote communication and social behavior in plants, helping them to establish beneficial relation-ships with other plants and organisms.
4. 741 Hz – This frequency is believed to promote overall plant health and resistance to disease.
5. 852 Hz – This frequency is believed to promote spiritual growth and connection with the natural world, helping plants to achieve their full

potential.

It is important to note that the effects of these frequencies on plant growth and health are still the subject of ongoing research and debate, and more studies are needed to fully understand their impact.

some facts

1. Plants are capable of complex communication with other plants and even with animals.
2. Plants can detect changes in their environment, such as changes in temperature, humidity, and light.
3. Plants can respond to sound and music, with certain frequencies and rhythms appearing to have a positive effect on plant growth and health.
4. Plants have a complex network of sensors and signaling pathways, which allow them to adapt to their environment.
5. Some plants are capable of releasing chemicals that can signal to other plants in their vicinity.
6. Plants can recognize and respond to the presence of predators and other threats.
7. Some plants can even recognize and respond to the presence of humans, including the sound of our footsteps.
8. Plants are capable of learning and memory, and can remember past events and experiences.
9. Plants can communicate with one another through underground networks of mycorrhizal fungi, which allow them to share nutrients and information.
10. Plants are capable of sensing and responding to changes in the weather and other environmental factors.
11. Some plants are able to move and adjust their position in response to changes in their environment.
12. Plants can produce chemicals that are used in medicine, including some of the most powerful and effective drugs.

13. Many traditional medicines are derived from plants, and are still used today to treat a variety of conditions.

14. Plants can be used to treat a wide range of medical conditions, from common colds to cancer.

15. Some plants are used to make fragrances, such as essential oils and perfumes.

16. Plants can be used to make natural dyes, which are used in clothing and textiles.

17. Some plants can be used to make biofuels, which are a renewable alternative to fossil fuels.

18. Plants can be used to clean up polluted environments, through a process known as phytoremediation.

19. Some plants are able to survive in extreme environments, such as deserts and arctic regions.

20. Some plants are able to survive for thousands of years, and may hold important clues to our understanding of the past.

21. Some plants are capable of producing electricity, through a process known as plant-electrodes.

22. Plants are capable of producing chemicals that can be used as insecticides, herbicides, and other types of pesticides.

23. Some plants are able to produce natural toxins that can be used to deter herbivores and other predators.

24. Plants are capable of adapting to changing conditions, and can evolve new traits over time.

25. Plants can be used to improve soil health, through a process known as phytoremediation.

26. Some plants are able to purify water, and can be used to treat polluted water sources.

27. Plants are capable of producing oxygen, which is essential for life on Earth.

28. Some plants are able to absorb and store carbon dioxide, which helps to reduce the amount of greenhouse gases in the atmosphere.

29. Plants play a crucial role in the Earth's ecosystem, and are responsible

for supporting all forms of life on the planet.

30. The study of plants continues to be a fascinating area of research, and holds important insights into the secrets of the natural world.

52

The Mystery of the Ginkgo Biloba

Ginkgo biloba, an ancient tree species native to China, has captivated botanists, horticulturists, and scientists for centuries due to its unique characteristics, resilience, and potential health benefits. Often referred to as a living fossil, the ginkgo tree has an enigmatic evolutionary history, with a lineage that stretches back over 270 million years. In this article, we will explore the mysterious origins of the ginkgo biloba, its unusual properties, and the ongoing research efforts aimed at understanding this remarkable plant.

A Living Fossil

Ginkgo biloba is the sole surviving species of the Ginkgophyta division, a group of ancient plants that once thrived across the globe. With fossils dating back to the Permian period, ginkgo trees have outlasted numerous mass extinction events, including the one that wiped out the dinosaurs 65 million years ago. The ginkgo's ability to survive and adapt to changing environments makes it an invaluable resource for studying plant evolution and resilience.

Unique Characteristics

The ginkgo tree possesses several unique characteristics that set it apart from other plant species:

1. Distinctive fan-shaped leaves: Ginkgo biloba leaves have a unique

fan shape, with veins that radiate out from the base of the leaf. The distinctive leaves turn a striking golden yellow in the fall, making the ginkgo a popular ornamental tree in urban landscapes.

2. Dioecious reproduction: Unlike most trees, ginkgo trees are dioecious, meaning that individual trees are either male or female. Female ginkgo trees produce a fleshy, fruit-like seed that emits a pungent odor when ripe, often compared to the smell of rancid butter.

3. Resilience: Ginkgo trees are highly adaptable and resistant to disease and pollution, which has allowed them to thrive in urban environments. They are also known for their remarkable ability to withstand extreme conditions, such as the ginkgo trees that survived the atomic bombing of Hiroshima in 1945.

Mysterious Origins and Dispersal

The ginkgo's evolutionary history is shrouded in mystery, with many unanswered questions regarding its origins and dispersal. While ginkgo fossils have been found on every continent except Antarctica, the tree is native only to China. Theories suggest that the ginkgo's range was drastically reduced during the last ice age, with the surviving trees confined to a few isolated refuges in China.

It was not until the 18th century that the ginkgo tree was introduced to Europe and North America by botanists and plant collectors, where it quickly gained popularity as an ornamental tree and a subject of scientific fascination.

Potential Health Benefits and Ongoing Research

Ginkgo biloba has been used in traditional Chinese medicine for thousands of years, and modern research has begun to explore its potential health benefits. Ginkgo extract, derived from the tree's leaves, is believed to have antioxidant and anti-inflammatory properties and is often marketed as a supplement to improve cognitive function and circulation.

However, scientific studies on the efficacy of ginkgo supplements have yielded mixed results, with some studies showing modest benefits, while others have found no significant effect. Despite the inconclusive evidence, the ginkgo's potential health benefits and ancient history continue to intrigue researchers and consumers alike.

The ginkgo biloba is a living testament to the resilience and adaptability of the plant kingdom, with a mysterious past that offers a window into the Earth's ancient history.

53

The Power of Sound Healing

Sound healing, an ancient practice that utilizes the power of sound and vibration to promote wellness and balance in the body, has experienced a resurgence in recent years. Rooted in various cultural and spiritual traditions, sound healing is now being explored by modern science for its potential therapeutic effects on physical, mental, and emotional health. In this article, we will delve into the history of sound healing, discuss the principles behind this holistic approach, and examine the growing body of research supporting its therapeutic potential.

History of Sound Healing

The use of sound as a healing modality can be traced back thousands of years across various cultures and spiritual traditions. Ancient civilizations such as the Egyptians, Greeks, and Chinese recognized the power of sound and music to heal and maintain harmony within the body and the environment. Indigenous cultures, such as Native Americans and Aboriginal Australians, have also long used sound in their rituals, ceremonies, and healing practices. Instruments such as Tibetan singing bowls, gongs, drums, bells, and tuning forks have been employed to facilitate healing and induce meditative states.

Principles of Sound Healing

Sound healing is based on the premise that everything in the universe, including the human body, is composed of vibrating energy. When the body's natural frequencies become imbalanced or disrupted, this can lead to dissonance, illness, or disease. Sound healing aims to restore harmony and balance within the body by using specific vibrations and frequencies to stimulate the body's natural healing mechanisms.

There are several different methods and techniques used in sound healing, including:

1. Vocal toning: The use of the human voice to produce specific tones or sounds that resonate with different parts of the body.
2. Singing bowls and gongs: These instruments generate a rich, multi-layered sound that can induce deep relaxation and stimulate the body's natural healing processes.
3. Tuning forks: Tuning forks are struck and placed on specific points on the body to deliver precise vibrations that can help to balance the body's energy systems.
4. Binaural beats: This technique involves listening to two slightly different frequencies, one in each ear, which encourages the brain to produce a third frequency that corresponds to a desired brainwave state (e.g., relaxation or focus).

Scientific Research on Sound Healing

Modern science has begun to explore the therapeutic potential of sound healing, with a growing body of research supporting its effects on various aspects of health:

1. Stress reduction: Numerous studies have demonstrated that sound healing can significantly reduce stress and anxiety levels, as well as lower cortisol levels, the hormone associated with stress.
2. Pain management: Research has shown that sound healing can effectively alleviate both acute and chronic pain, making it a promising complementary therapy for pain management.

3. Sleep improvement: Sound healing has been found to improve sleep quality, with studies showing positive effects on insomnia and other sleep disorders.
4. Cognitive function: Preliminary research suggests that sound healing may have a positive impact on cognitive function, including memory and focus.
5. Emotional well-being: Sound healing has been shown to improve mood, reduce symptoms of depression, and promote feelings of relaxation and well-being.

The power of sound healing, an ancient practice rooted in various cultural and spiritual traditions, is being rediscovered and embraced in our modern world. As scientific research continues to validate its potential therapeutic benefits, sound healing is gaining recognition as a valuable complementary therapy for various health conditions. By harnessing the power of sound and vibration, sound healing offers a holistic approach to well-being that aligns the body, mind, and spirit, paving the way for optimal health and harmony.

The use of music for healing purposes dates back thousands of years, with many ancient cultures incorporating music and sound into their healing practices. In recent years, there has been growing interest in the use of specific frequencies, known as solfeggio frequencies, for promoting healing and well-being.

Solfeggio frequencies are a series of specific tones that are believed to have a special relationship with the chakras, or energy centers, in the body. When these tones are played together, they create a powerful healing vibration that can help to balance and align the chakras and develop a meditative mind.

There are several frequencies that fall into this category, including 174 Hz, 285 Hz, 396 Hz, 417 Hz, 432 Hz, 528 Hz, 639 Hz, 741 Hz, and 852 Hz. Each of these frequencies is believed to have unique healing properties, and can be used to address a variety of physical and emotional issues.

Research has shown that exposure to solfeggio tones can have a number of health benefits. For example, studies have found that the 528 Hz frequency can reduce stress hormones in the body, while sounds at 432 Hz have been shown to lower blood pressure and breathing rates and promote feelings of satisfaction.

Sound healing can also be a natural anesthetic, as it can help to relieve pain. One study found that sounds at 432 Hz are beneficial in reducing dental anxiety and relieving pain during dental surgery.

In addition to its physical benefits, solfeggio frequency music and sound healing can also have a positive impact on mental and emotional health. Many people report feeling deep relaxation, increased feelings of well-being, and a soothing of negative emotions when listening to solfeggio tones. Sound healing has also been shown to improve sleep quality, heighten clarity and intuition, and promote faster healing from injuries.

While the use of solfeggio frequencies and sound healing is still the subject of ongoing research and debate, many people have reported positive experiences and benefits from incorporating these practices into their lives. Whether you are seeking relief from physical pain, emotional stress, or simply looking to enhance your overall sense of well-being, the power of sound and music can be a powerful tool for healing and transformation.

54

The Secrets of Sacred Geometry

For centuries, many ancient cultures have used sacred geometry as a means of unlocking the mysteries of the universe. Sacred geometry refers to the belief that certain geometric shapes and patterns hold powerful spiritual and energetic properties. These shapes and patterns can be found in the natural world, in the structures of our bodies, and in many of the most sacred and revered religious and spiritual symbols.

One of the most widely recognized sacred geometry patterns is the flower of life, which consists of a series of overlapping circles that form a hexagonal pattern. This pattern can be found in many different cultures and spiritual traditions, from ancient Egypt to Hinduism to modern-day New Age beliefs.

Another common sacred geometry shape is the vesica piscis, which is created by the intersection of two circles of the same size. This shape has been used in many different cultures to represent the divine feminine and masculine energies, and is often associated with the balance and harmony of the universe.

The golden ratio is another important concept in sacred geometry, and is often referred to as the divine proportion. This ratio, which is approximately 1.618, is believed to be found throughout the natural world, from the patterns of flower petals to the spiral of a seashell.

Many people believe that by meditating on these sacred geometry patterns and shapes, we can tap into the universal energy and unlock deeper levels of spiritual understanding. Some also believe that by using these shapes in our physical surroundings, such as in the design of buildings or in the layout of gardens, we can create spaces that are in harmony with the natural world and conducive to spiritual growth and well-being.

In recent years, there has been growing interest in the use of sacred geometry in art and design, with many artists and designers incorporating these patterns and shapes into their work. Some even believe that by surrounding ourselves with these sacred geometric forms, we can improve our physical, emotional, and spiritual health.

While the use of sacred geometry is still the subject of ongoing research and debate, many people have reported positive experiences and benefits from incorporating these practices into their lives. Whether you are seeking a deeper connection to the universe, or simply looking to create a more harmonious and balanced environment for yourself, the secrets of sacred geometry may hold the key to unlocking your full potential and experiencing the transformative power of the natural world.

The golden ratio

The golden ratio is a mathematical ratio that is approximately 1.61803398875. It is also known as the golden mean or the divine proportion, and it has been found in many natural and man-made structures throughout history. The secret of the golden ratio lies in its aesthetic appeal and its ability to create balance and harmony in design.

One of the most famous examples of the golden ratio is the Fibonacci sequence, which is a series of numbers where each number is the sum of the two preceding numbers. When plotted on a graph, these numbers create a spiral that approximates the golden ratio. This spiral can be found in many natural phenomena, such as the growth patterns of shells, pine cones, and sunflowers.

The golden ratio has been used in art and design for centuries, with examples found in the works of Leonardo da Vinci, Michelangelo, and many other famous artists. It is believed that the use of the golden ratio creates a sense of balance and proportion that is pleasing to the eye.

The golden ratio has also been used in architecture, with examples found in the designs of the Parthenon in Greece and the Great Mosque of Kairouan in Tunisia. In these structures, the golden ratio is used to create a sense of harmony and balance between different elements of the design.

The golden ratio has also been found in music, with composers such as Mozart and Beethoven using it to create harmonious melodies and rhythms. In fact, it has been suggested that the use of the golden ratio in music can create a sense of emotional resonance with the listener.

In modern times, the golden ratio has been applied to many different fields, such as graphic design, product design, and even website design. It is believed that the use of the golden ratio can create a sense of balance and proportion that is appealing to consumers and can help to enhance the overall user experience.

Overall, the secret of the golden ratio lies in its ability to create harmony and balance in design. Whether it is found in nature, art, or architecture, the golden ratio has an aesthetic appeal that has fascinated people for centuries and continues to inspire creativity and innovation today.

Here Some Facts

1. Sacred geometry is the study of geometric patterns and principles that are considered to have spiritual or mystical significance.
2. Sacred geometry has been used by many cultures throughout history, including ancient Egypt, Greece, and India.
3. Sacred geometry is based on the belief that certain shapes and patterns have inherent meanings and energies that can be used for healing and spiritual development.
4. Some of the most common shapes used in sacred geometry include the circle, triangle, square, and pentagon.
5. The golden ratio, a mathematical proportion found in nature and art,

is often considered to be a sacred geometric principle.

6. The Flower of Life, a complex geometric pattern consisting of overlapping circles, is a popular symbol in sacred geometry.

7. The study of sacred geometry can be used to understand the underlying patterns and structures of the universe.

8. Sacred geometry is often used in architecture and design to create spaces that are in harmony with the natural environment.

9. The study of sacred geometry has been used in spiritual practices such as meditation and visualization to promote healing and spiritual growth.

10. Sacred geometry can be used to create mandalas, intricate patterns used in Hindu and Buddhist practices to represent the universe and promote inner harmony.

11. Some proponents of sacred geometry believe that it can be used to access higher states of consciousness and spiritual realms.

12. Sacred geometry has been used in crop circle formations, leading some to believe that they are created by extraterrestrial or supernatural beings.

13. Some researchers have studied the use of sacred geometry in ancient cultures to gain insights into their spiritual beliefs and practices.

14. Sacred geometry has been used in music and sound therapy to create harmonious and healing vibrations.

15. The study of sacred geometry can be used to create art that is imbued with spiritual significance and meaning.

16. The principles of sacred geometry have been incorporated into various new age and spiritual movements.

17. The study of sacred geometry continues to evolve and adapt to modern times, with new techniques and applications being developed and explored.

18. Sacred geometry is often associated with the concept of universal consciousness and the interconnectedness of all things.

19. The study of sacred geometry can be used to promote mindfulness and spiritual awareness in daily life.

20. The study of sacred geometry can be a powerful tool for self-discovery and personal growth, helping individuals to connect with their higher selves and the world around them.

55

The Science of Astrology

Astrology, the study of the movements and relative positions of celestial objects and their influence on human affairs, has been a subject of fascination
and debate for thousands of years. Rooted in ancient civilizations, astrology
has evolved into a complex system that seeks to provide insight into
human nature, relationships, and future events. While often dismissed as
pseudoscience by the scientific community, astrology continues to captivate
the imagination of millions who seek guidance and understanding through
the lens of the cosmos. In this article, we will explore the history of astrology,
the principles that underpin this intriguing discipline, and the ongoing
debate surrounding its validity.

History of Astrology

The origins of astrology can be traced back to the ancient civilizations of
Mesopotamia, Egypt, China, and India, where the movements of celestial
bodies were meticulously observed and recorded. The Babylonians are
credited with the development of the zodiac, the twelve constellations that
form the backdrop for the apparent path of the sun, moon, and planets. As
astrology spread throughout the ancient world, it was embraced and adapted
by various cultures, leading to the development of distinct systems such as
Western, Vedic, and Chinese astrology.

Principles of Astrology

Astrology is based on the premise that the positions and movements of celestial bodies can provide insight into human personality traits, relation-ships, and events. The main components of astrology include:

1. The zodiac: The zodiac is comprised of twelve constellations, each associated with specific personality traits and qualities. A person's sun sign, determined by the position of the sun at the time of birth, is the most well-known aspect of astrology and is said to influence one's core identity.

2. The planets: In astrology, each planet represents different aspects of human experience, such as communication, love, and ambition. The positions of the planets in relation to the zodiac signs at the time of birth are believed to shape an individual's personality and life experiences.

3. The houses: The astrological chart is divided into twelve houses, each representing a specific area of life, such as career, relationships, and health. The placement of planets within these houses provides insight into how the planetary energies manifest in various aspects of a person's life.

4. Aspects: Aspects are the angles formed between planets in the astro-logical chart, which are said to reveal the nature of the relationships between the planets and their influence on an individual's life.

The Debate Surrounding Astrology

The validity of astrology has long been a subject of debate and skepticism within the scientific community. Critics argue that there is no empirical evidence to support the claims made by astrology and that any perceived accuracy can be attributed to the Barnum effect, a cognitive bias that causes people to perceive general statements as being highly accurate and personal.

However, proponents of astrology argue that the discipline provides valuable insights into human nature and that the connection between celestial events and human affairs is not yet fully understood. While the scientific community remains largely unconvinced by astrology's claims, interest in the subject persists, with many people finding value in the personal insights and guidance that astrology can offer.

Astrology, a discipline with roots in ancient civilizations, continues to captivate the imagination of millions who seek to understand the connection between the cosmos and human life. While the scientific community largely dismisses astrology as a pseudoscience, its enduring appeal suggests that the quest for self-understanding and cosmic connection remains a deeply ingrained aspect of the human experience.

some general characteristics associated with each zodiac sign:

Aries (March 21 - April 19): Known for their confidence, courage, and leadership skills, Aries tend to be adventurous and always seeking new challenges. They can also be impulsive and may struggle with patience.

Taurus (April 20 - May 20): Taureans are often practical, reliable, and hardworking. They value stability and security, and may struggle with change or uncertainty. They are also known for their stubbornness.

Gemini (May 21 - June 20): Geminis are often quick-witted, curious, and adaptable. They tend to be great communicators and are always seeking new experiences. However, they may struggle with consistency and may have a tendency towards indecisiveness.

Cancer (June 21 - July 22): Cancerians tend to be nurturing, empathetic, and sensitive. They value emotional connections and may be prone to mood swings. They can also be quite protective of themselves and their loved ones.

Leo (July 23 - August 22): Leos are often confident, creative, and enthusiastic. They tend to be natural leaders and enjoy being the center of attention. However, they may struggle with criticism or rejection.

Virgo (August 23 - September 22): Virgos tend to be analytical, practical, and detail-oriented. They value organization and structure, and may be perfectionistic. They can also be quite critical of themselves and others.

Libra (September 23 - October 22): Libras are often diplomatic, charming, and cooperative. They value harmony and balance, and may struggle with conflict. They can also be indecisive and may struggle with setting boundaries.

Scorpio (October 23 - November 21): Scorpios tend to be intense, passionate, and mysterious. They value honesty and authenticity, and may struggle with trust issues. They can also be quite possessive and may have a tendency towards jealousy.

Sagittarius (November 22 - December 21): Sagittarians are often adventurous, optimistic, and philosophical. They tend to have a love for travel and exploration, and may struggle with commitments or routine. They can also be blunt and may struggle with tact.

Capricorn (December 22 - January 19): Capricorns tend to be ambitious, responsible, and disciplined. They value hard work and success, and may struggle with taking risks. They can also be quite reserved and may struggle with vulnerability.

Aquarius (January 20 - February 18): Aquarians are often independent, innovative, and humanitarian. They tend to value equality and social justice, and may struggle with conformity. They can also be quite eccentric and may struggle with emotional intimacy.

Pisces (February 19 - March 20): Pisceans tend to be compassionate, artistic, and intuitive. They value emotional connections and may be quite empathetic. They can also be quite sensitive and may struggle with boundaries.

It is important to remember that these are generalizations and not every person born under a specific zodiac sign will exhibit these traits. Additionally, astrology should not be used to make assumptions about individuals or to categorize them into specific personality types.

56

The Enigma of Crop Circles: Natural Phenomenon or Alien Signaling?

Crop circles, intricate and large-scale patterns that appear in fields of crops overnight, have captured the public's imagination and sparked heated debates among scientists, paranormal enthusiasts, and skeptics since their rise to prominence in the late 20th century. These elaborate formations have been attributed to various causes, ranging from natural phenomena and human-made hoaxes to extraterrestrial communication. In this article, we will delve into the mystery of crop circles, examine the different theories surrounding their origins, and explore the ongoing investigations into this enigmatic phenomenon.

The History of Crop Circles

Crop circles have been reported throughout history, with some accounts dating back to the 17th century. However, it wasn't until the 1970s and 1980s that the phenomenon gained widespread attention, with an increasing number of complex and elaborate formations appearing in fields across the United Kingdom and other countries.

Theories Behind Crop Circles

There are several competing theories about the origin and purpose of crop circles:

1. Natural phenomena: Some researchers believe that crop circles may be the result of natural phenomena, such as unusual weather patterns, wind vortices, or ball lightning. These theories propose that these natural forces could create the intricate patterns seen in crop circles by flattening the crops in a specific manner.

2. Human-made hoaxes: Skeptics argue that crop circles are nothing more than elaborate hoaxes created by humans for fun, artistic expression, or to generate publicity. Indeed, some individuals and groups have come forward, claiming responsibility for creating specific crop circles using simple tools such as ropes and wooden boards.

3. Extraterrestrial communication: The most controversial theory posits that crop circles are the work of extraterrestrial beings attempting to communicate with humans or mark their presence on Earth. Proponents of this theory argue that the complexity, precision, and scale of some crop circles are beyond human capabilities and must be the result of advanced technology or intelligence.

Ongoing Investigations

Despite the myriad of theories surrounding crop circles, there is no consensus among researchers on their origins. Investigations into the phenomenon continue, with some scientists studying the physical and biological effects on the crops and soil within crop circles, while others focus on the mathematical and geometric properties of the formations.

For instance, researchers have discovered that some crop circles exhibit unusual electromagnetic properties and alterations in the cellular structure of the plants, suggesting that an unknown force or energy may be involved in their creation. Others have noted the geometric precision and complex designs found in some crop circles, which seem to incorporate mathematical concepts and ancient symbols.

The Cultural Impact of Crop Circles

Crop circles have had a significant impact on popular culture, inspiring countless books, documentaries, movies, and art installations. They have also generated a thriving subculture of enthusiasts known as "croppies," who visit and document crop circle formations, organize conferences, and engage in spirited debates about their origins and meaning.

The enigma of crop circles continues to fascinate and perplex researchers, enthusiasts, and skeptics alike. As investigations into this mysterious phenomenon persist, the debate over their origins – whether natural, human-made, or extraterrestrial – remains unresolved. Regardless of their true source, crop circles have undoubtedly captured the public's imagination, serving as a powerful symbol of the enduring quest for understanding and connection with the unknown.

57

The Secret of Animal Migration: How Do They Navigate the World?

Animal migration is a fascinating phenomenon that has long puzzled scientists and animal enthusiasts alike. Every year, millions of animals travel vast distances across the globe, navigating through a variety of different environments and weather conditions to reach their final destination. But how do they do it? What is the secret to animal migration?

There are many different factors that contribute to animal migration, including changes in food availability, mating opportunities, and weather patterns. But one of the key factors that allows animals to successfully navigate their way across vast distances is their ability to sense the Earth's magnetic field.

Many migratory animals, such as birds, sea turtles, and certain fish species, have magnetoreceptors in their bodies that allow them to detect the Earth's magnetic field. This is believed to help them navigate and orient themselves as they travel, using the magnetic field as a kind of compass.

Research has shown that migratory birds, for example, are able to sense the Earth's magnetic field through specialized cells located in their eyes, beaks, and inner ears. These cells contain tiny particles of magnetite, a magnetic mineral that allows the birds to detect and interpret changes in the magnetic field as they travel.

Other animals, such as monarch butterflies, use a combination of different cues to navigate during migration. Monarchs are able to detect changes in light and temperature, as well as the position of the sun, to help guide their migration.

While the mechanisms of animal migration are still not fully understood, research into the field is ongoing. Scientists are using a variety of techniques, including GPS tracking and genetic analysis, to better understand how animals navigate and orient themselves during migration.

Understanding the secret of animal migration is not only fascinating from a scientific perspective, but it also has important implications for conservation efforts. By understanding how migratory animals navigate and travel, we can work to better protect their habitats and ensure their survival for generations to come.

Facts

1. Animal migration is a natural phenomenon that has been observed in a wide range of species, from birds and mammals to fish and insects.
2. Many animals migrate in search of food, water, or better breeding opportunities.
3. Some species migrate thousands of miles across different continents and oceans, while others migrate shorter distances within their local regions.
4. Migration patterns can vary greatly between different species, with some animals migrating in a straight line and others following specific routes.
5. Animal migration can occur on a seasonal or annual basis, depending on the species and their individual needs.
6. One of the key factors that allows animals to navigate during migration is their ability to sense the Earth's magnetic field.
7. Many migratory animals have specialized cells in their bodies that

contain magnetic minerals, such as magnetite, which allow them to detect changes in the Earth's magnetic field.

8. Other animals, such as monarch butterflies, use visual cues such as the position of the sun and landmarks to navigate during migration.

9. The length of an animal's migratory journey can vary greatly depending on the species and the distance they need to travel.

10. Migration can be a dangerous process for animals, as they may face threats from predators, extreme weather conditions, and changes in their environment.

11. Migration is a costly process for animals, as it requires a lot of energy and can put them at risk of injury or death.

12. Many migratory species are facing threats to their survival, including habitat loss, climate change, and hunting.

13. Scientists are studying animal migration to better understand the mechanisms behind it, and to develop conservation strategies to protect migratory species.

14. GPS tracking and satellite imagery are some of the tools that researchers use to track and monitor animal migration.

15. Many migratory animals travel in large groups, such as flocks of birds or schools of fish, which can help protect them from predators.

16. Some species of salmon are able to migrate upstream to spawn in the same river where they were born, using their sense of smell to navigate.

17. Certain species of sea turtles travel thousands of miles to reach their breeding grounds, using the Earth's magnetic field to guide them.

18. Some species of whales migrate great distances each year to feed in colder waters during the summer months.

19. The timing of migration can be affected by environmental factors such as temperature and daylight hours.

20. Many migratory species have complex social structures and communication systems that play a role in their migration patterns.

58

The Mystery of the Loch Ness Monster: Fact or Fiction?

The Loch Ness Monster, affectionately known as "Nessie," is a legendary creature said to inhabit the deep, murky waters of Scotland's Loch Ness. For centuries, tales of a large, serpent-like creature have captivated the imaginations of locals and visitors alike, with numerous sightings, photographs, and expeditions adding fuel to the mystery. Despite the lack of definitive evidence, the legend of the Loch Ness Monster endures as one of the world's most famous unsolved mysteries. In this article, we will explore the history of Nessie, examine the various theories and explanations for the sightings, and discuss the ongoing search for the elusive creature.

History of the Loch Ness Monster

While stories of water-dwelling creatures can be traced back to ancient Celtic mythology, the modern legend of the Loch Ness Monster began in the 1930s with a series of reported sightings and the publication of the now-iconic "Surgeon's Photograph." This photograph, purportedly showing the head and neck of Nessie emerging from the water, ignited a media frenzy and cemented the Loch Ness Monster's status as a global phenomenon.

Theories and Explanations

Numerous theories have been proposed to explain the sightings and stories of the Loch Ness Monster, ranging from misidentifications of common animals to elaborate hoaxes:

1. Misidentification of known animals: Some researchers believe that sightings of the Loch Ness Monster can be attributed to misidentifications of known animals, such as large fish, seals, or even birds. The lake's murky waters, low visibility, and strong currents could create optical illusions or cause natural objects to appear unusual or mysterious.

2. Hoaxes and pranks: Skeptics argue that many of the alleged sightings and photographs of Nessie are hoaxes, created for fun or financial gain. The "Surgeon's Photograph," for example, was later revealed to be a staged hoax using a toy submarine and a model head and neck.

3. Prehistoric creatures: One of the more popular theories among Nessie enthusiasts is that the Loch Ness Monster is a surviving member of an ancient species, such as a plesiosaur or an eel-like creature. Proponents of this theory argue that Loch Ness, with its vast size, depth, and numerous underwater caves, could provide a suitable habitat for such a creature to survive undetected.

The Search for Nessie

Over the years, numerous expeditions and scientific investigations have attempted to uncover evidence of the Loch Ness Monster. These efforts have included sonar scans, underwater photography, and even satellite imaging. Despite these extensive searches, no definitive evidence of Nessie's existence has been found.

However, these investigations have not deterred Nessie enthusiasts, who continue to search for the elusive creature and analyze new sightings and photographs. Some researchers have even turned to advanced technologies such as environmental DNA (eDNA) analysis, which can detect the presence of organisms by analyzing the genetic material left behind in the water.

Cultural Impact

The Loch Ness Monster has had a profound impact on popular culture, inspiring countless books, movies, TV shows, and documentaries. The legend of Nessie has also transformed Loch Ness into a major tourist destination, attracting visitors from around the world who hope to catch a glimpse of the mysterious creature.

The mystery of the Loch Ness Monster remains one of the world's most enduring unsolved enigmas.

59

The Power of Natural Remedies: The Healing Properties of Plants and Herbs

For centuries, people have used natural remedies and traditional medicines to treat a variety of health conditions. These remedies, which are often derived from plants and herbs, have been shown to have powerful healing properties and can be an effective alternative to modern medicine in certain situations.

Plants and herbs contain a wide range of compounds that can have medicinal benefits. For example, chamomile is often used to treat insomnia and anxiety, while ginger has been shown to be effective in reducing nausea and vomiting. Other popular remedies include peppermint, which is often used to soothe digestive issues, and lavender, which is used to promote relaxation and reduce stress.

One of the benefits of natural remedies is that they are often much gentler on the body than prescription medications. Unlike synthetic drugs, which can have a range of side effects, natural remedies are generally well-tolerated and are less likely to cause harm.

Another benefit of natural remedies is that they can be more accessible and affordable than traditional medicines. Many common remedies can be grown in a home garden or purchased at a local health food store, making them a convenient and cost-effective option for people looking to improve their health.

Despite their many benefits, it is important to note that natural remedies are not a substitute for modern medical treatments. While they can be effective for treating certain conditions, they are not appropriate for all health concerns and may not be as effective as prescription medications in some cases.

Additionally, it is important to be cautious when using natural remedies, as they can interact with other medications and may not be safe for everyone. It is always best to consult with a healthcare provider before using any new remedy, especially if you are pregnant or have a chronic medical condition.

Overall, the power of natural remedies and the healing properties of plants and herbs should not be underestimated. With proper use and caution, they can be a safe and effective way to promote health and wellbeing.

Here is a list of some common herbs and their potential medicinal properties:

1. Chamomile - may help with insomnia and anxiety
2. Echinacea - may boost the immune system and help prevent colds and flu
3. Ginger - may help reduce nausea and vomiting
4. Peppermint - may soothe digestive issues and relieve headaches
5. Turmeric - may have anti-inflammatory properties and help reduce pain
6. Lavender - may promote relaxation and reduce stress
7. St. John's Wort - may help with mild to moderate depression and anxiety
8. Garlic - may help lower cholesterol and blood pressure
9. Rosemary - may improve memory and concentration
10. Sage - may help with memory and cognitive function
11. Thyme - may have antibacterial and antifungal properties
12. Aloe vera - may help soothe skin irritations and promote wound healing
13. Calendula - may help reduce inflammation and promote wound healing
14. Dandelion - may help improve digestion and liver function

15. Milk thistle – may help protect the liver and improve liver function
16. Valerian – may help with insomnia and anxiety
17. Passionflower – may help with anxiety and improve sleep quality
18. Lemon balm – may help reduce anxiety and improve cognitive function
19. Ginkgo biloba – may improve cognitive function and memory
20. Ashwagandha – may help reduce stress and anxiety and improve mood.

60

The Secrets of Geomancy: The Study of Earth's Energies.

Geomancy is the ancient art and practice of divination that is based on interpreting the patterns and formations found in the natural landscape. It is a method of accessing the earth's energies and using them for healing, protection, and guidance. The study of geomancy can reveal many secrets about the world we live in and how we interact with it.

One of the main principles of geomancy is the belief that the earth is alive and has its own consciousness. This consciousness is said to be expressed through the patterns and formations that can be found in the landscape, such as the shape of mountains, the flow of rivers, and the patterns of rocks and crystals.

Geomancy is often used to identify areas of land that are particularly powerful or significant, such as ley lines or vortexes. These areas are believed to be points of high energy and are often associated with sacred sites, such as Stonehenge or Machu Picchu.

One of the tools used in geomancy is the dowsing rod, which is used to locate water, minerals, and other natural resources. The rod is said to be able to detect subtle changes in the earth's energy field, allowing the dowser to locate underground streams, mineral deposits, and other hidden resources.

Another aspect of geomancy is the study of feng shui, which is a Chinese practice that focuses on the placement of objects and structures in the environment in order to achieve balance and harmony. Feng shui is based on the belief that the energy, or chi, that flows through the environment can affect our health, happiness, and success.

Geomancy can also be used for healing purposes, such as in the practice of crystal healing. Crystals are believed to have their own unique energy and can be used to balance the body's energy field, promote healing, and enhance spiritual development.

The study of geomancy can provide insights into the natural world and our place within it. By understanding the earth's energies and how they interact with our own, we can develop a deeper connection to the environment and gain a greater appreciation for the interconnectedness of all things.

Some Facts

1. Geomancy is an ancient practice that has been used by cultures around the world for thousands of years.
2. The word "geomancy" comes from the Greek words for "earth" and "divination."
3. The practice of geomancy is based on the belief that the earth has its own consciousness and energy field.
4. Geomancers use various tools, such as dowsing rods and pendulums, to detect and measure the earth's energies.
5. The patterns and formations found in the natural landscape, such as mountains, rivers, and rocks, are considered to be expressions of the earth's consciousness and energy.
6. Geomancy is often used to identify areas of land that are particularly powerful or significant, such as ley lines or vortexes.
7. Feng shui, a Chinese practice that focuses on the placement of objects and structures in the environment to achieve balance and harmony, is a form of geomancy.
8. The study of geomancy can provide insights into the natural world and our place within it.

9. Geomancers believe that the earth's energies can affect our health, happiness, and success.

10. The practice of crystal healing, which involves using crystals to balance the body's energy field, is based on the principles of geomancy.

11. Geomancers believe that the earth's energies can be used for healing purposes, such as in the practice of earth acupuncture.

12. The use of dowsing rods to locate underground water sources and minerals is a form of geomancy.

13. Geomancers may use astrology and other divinatory tools to gain insights into the earth's energies and their effects on individuals and society.

14. Some geomancers believe that the earth's energies can be used to predict natural disasters and other major events.

15. The study of geomancy can be used to design buildings and other structures that are in harmony with the natural environment.

16. Geomancy is often associated with spiritual and esoteric practices, such as shamanism and alchemy.

17. The practice of geomancy has been criticized by some scientists and skeptics who view it as pseudoscience.

18. The use of feng shui in architecture and interior design has become increasingly popular in recent years.

19. The principles of geomancy have been incorporated into various spiritual and new age movements.

20. The study of geomancy continues to evolve and adapt to modern times, with new techniques and practices being developed and explored.

VII

Secrets of the Human Mind

61

The Science of Dreams

Dreams have been a source of fascination and mystery for centuries. They can be vivid, surreal, and often nonsensical, leaving us with a feeling of both wonder and confusion. While many theories about the purpose of dreams have been proposed, the science of dreams is still a topic of ongoing research and debate. In this article, we will explore the science of dreams, including the different stages of sleep, the neuroscience behind dreams, and the potential benefits and drawbacks of dreaming.

The Stages of Sleep

To understand the science of dreams, it is important to first understand the different stages of sleep. Sleep is divided into two main categories: rapid eye movement (REM) sleep and non-rapid eye movement (NREM) sleep.

During NREM sleep, the brainwaves slow down, and the body relaxes. This stage of sleep is divided into three sub-stages, with each stage becoming progressively deeper.

During REM sleep, the brain is highly active, and the eyes move rapidly. This stage of sleep is when most dreaming occurs, and it is often associated with vivid, emotionally intense dreams.

The Neuroscience of Dreams

While the exact purpose of dreaming is still a topic of debate, neuroscientists have made significant progress in understanding the neuroscience behind dreams.

One theory suggests that dreams are a form of mental processing, allowing the brain to consolidate memories and process emotions. During REM sleep, the brain is highly active, and it may be working to integrate new information with existing knowledge and memories.

Another theory suggests that dreams may be a form of problem-solving. During sleep, the brain is free from the distractions and biases of waking life, allowing it to approach problems from a new perspective and generate creative solutions.

The Benefits and Drawbacks of Dreaming

While dreaming may seem like a purely subjective experience, it may have both benefits and drawbacks for our overall health and well-being.

On the positive side, dreaming may help us to process emotions and consolidate memories. It may also promote creative problem-solving and help us to generate new insights and ideas.

However, dreaming can also have drawbacks, such as disrupting sleep quality and causing nightmares or anxiety. Dreams can also be influenced by external factors, such as stress or trauma, which can impact our mental health and well-being.

The science of dreams is a complex and evolving field, with many unanswered questions and ongoing research. While the exact purpose of dreams remains a mystery, there is evidence to suggest that they play an important role in mental processing and problem-solving.

Understanding the different stages of sleep and the neuroscience behind dreams can help us to better understand the nature of dreaming and its potential benefits and drawbacks. As research in this field continues to progress, we may gain new insights into the purpose of dreams and their impact on our overall health and well-being.

62

The Mystery of Hypnosis

Hypnosis is a state of focused attention and suggestibility that has fascinated scientists, psychologists, and the general public for centuries. Despite decades of research, the nature of hypnosis remains a mystery, with no consensus on its underlying mechanisms or effectiveness. In this article, we will explore the mystery of hypnosis, including its history, theories, and potential applications.

The History of Hypnosis

The use of hypnosis can be traced back to ancient times, with evidence of hypnotic-like practices in ancient Egypt and Greece. However, it was not until the 18th century that hypnosis began to be recognized as a distinct field of study, with the work of Franz Mesmer, a German physician who developed a theory of animal magnetism and claimed to induce a trance-like state in his patients.

In the 19th century, hypnosis became more widely recognized as a legitimate medical practice, with the work of James Braid, a Scottish physician who coined the term "hypnosis" and developed a more scientific approach to the study of hypnosis.

Theories of Hypnosis

Despite centuries of research, there is no consensus on the underlying mechanisms of hypnosis. Many theories have been proposed, including the role of suggestibility, the power of the hypnotist, and the activation of specific brain regions.

One popular theory of hypnosis suggests that it is a state of heightened suggestibility, in which the subject is more receptive to suggestions and more likely to experience alterations in perception, memory, and behavior.

Another theory proposes that hypnosis is a form of dissociation, in which the subject's conscious awareness is separated from their normal experience of reality. This theory suggests that hypnosis may be similar to other forms of dissociation, such as daydreaming or absorption in a book or movie.

Applications of Hypnosis

Despite the ongoing debate over its underlying mechanisms, hypnosis has been used in a variety of clinical and therapeutic settings, including pain management, anxiety reduction, and behavior modification.

In the field of pain management, hypnosis has been shown to be effective in reducing the perception of pain, as well as the need for pain medication. In the treatment of anxiety disorders, hypnosis has been used to help patients develop coping strategies and reduce symptoms.

Hypnosis has also been used in the field of behavior modification, with studies showing that it can be effective in helping people quit smoking, lose weight, and overcome phobias.

The mystery of hypnosis continues to fascinate scientists, psychologists, and the general public. While there is no consensus on the underlying mechanisms of hypnosis, there is evidence to suggest that it can be an effective tool in the treatment of a variety of clinical and therapeutic conditions.

As research in this field continues to progress, we may gain new insights into the nature of hypnosis and its potential applications.

63

The Secrets of Telepathy

Telepathy, the ability to communicate through means other than speech or writing, has long been a subject of fascination and skepticism. While many people claim to have experienced telepathic communication or have witnessed it, the scientific community remains divided on the existence and nature of telepathy. In this article, we will explore the secrets of telepathy, including its history, theories, and potential applications.

The History of Telepathy

The concept of telepathy can be traced back to ancient times, with references to telepathic communication found in various cultures and religions. However, it was not until the 19th century that telepathy began to be studied as a distinct phenomenon, with the work of scientists such as Frederic Myers and Charles Richet.

During the early 20th century, telepathy became a popular subject of study in the field of parapsychology, with researchers conducting experiments to test the existence and nature of telepathic communication.

Theories of Telepathy

Telepathy, the ability to communicate through means other than speech or writing, has long been a subject of fascination and skepticism. Despite the ongoing debate on its existence and nature, many theories have been proposed to explain telepathy. In this article, we will explore some of the most prominent theories of telepathy.

· **Extrasensory Perception** (ESP)

One of the most popular theories of telepathy suggests that it is a form of extrasensory perception (ESP). According to this theory, the sender and receiver are able to connect through a shared consciousness or field of energy. This field of energy is said to exist beyond the physical realm, allowing for the transmission of thoughts and feelings between individuals.

Proponents of this theory point to the many documented cases of telepathic communication, including instances of individuals who have been able to accurately predict the thoughts and actions of others. However, skeptics argue that these cases can often be explained by other means, such as intuition or chance.

· **Mental Projection**

Another theory of telepathy proposes that it is a form of mental projection, in which the sender is able to project thoughts or images into the mind of the receiver. According to this theory, telepathic communication is possible because the sender is able to create a mental image or thought that is powerful enough to be perceived by the receiver.

Proponents of this theory point to the many cases of telepathic communication that have been reported by individuals who claim to have projected their thoughts or images into the minds of others. However, skeptics argue that this theory is difficult to test and verify scientifically.

· **Quantum Entanglement**

A more recent theory of telepathy suggests that it may be linked to the phenomenon of quantum entanglement. According to this theory, two particles can become entangled, meaning that a change in the state of one particle will cause a corresponding change in the state of the other, regardless of the distance between them.

Proponents of this theory argue that telepathy may be a form of quantum entanglement, allowing for the transmission of thoughts and feelings between individuals without the need for physical contact. However, skeptics point out that the phenomenon of quantum entanglement is still poorly understood and has not been proven to occur at the level of the human brain.

Despite decades of research, the nature of telepathy remains a mystery, with no consensus on its underlying mechanisms or effectiveness. While many theories have been proposed, including the role of extrasensory perception (ESP), mental projection, and quantum entanglement, none have been scientifically proven.

As research in this field continues to progress, we may gain new insights into the nature of telepathy and its potential applications. Until then, telepathy remains a subject of fascination and skepticism, shrouded in mystery and intrigue.

Applications of Telepathy

While the scientific community remains divided on the existence and nature of telepathy, there are many who believe in its potential applications. Telepathic communication could have practical applications in a variety of fields, including communication, education, and healthcare.

In the field of communication, telepathy could allow for instant and seamless communication between individuals, without the need for speech or writing. In education, telepathic communication could revolutionize the way we learn and teach, allowing for instant transmission of knowledge and information.

In healthcare, telepathic communication could be used to diagnose and treat a variety of conditions, such as mental illness or neurological disorders. It could also be used to help individuals with disabilities, such as those who are nonverbal or have difficulty communicating through traditional means.

The secrets of telepathy continue to elude the scientific community, with no consensus on its underlying mechanisms or existence. While many people claim to have experienced telepathic communication or have witnessed it, the scientific community remains skeptical.

Despite this, there are many who believe in the potential applications of telepathy, including in communication, education, and healthcare. As research in this field continues to progress, we may gain new insights into the nature of telepathy and its potential applications.

64

The Enigma of Near-Death Experiences

Near-death experiences (NDEs) have been described as a mystical and transformative experience that occurs when a person is on the brink of death. These experiences have been reported across different cultures and religions, and often involve a sense of leaving one's body, seeing a bright light, and feeling a sense of peace or transcendence. Despite decades of research, the nature and underlying mechanisms of NDEs remain a mystery. In this article, we will explore the enigma of near-death experiences.

The Phenomenon of NDEs

Near-death experiences have been reported by people of all ages, cultures, and backgrounds. Many people describe feeling as if they are floating above their bodies and watching events from a distance. Others report a sense of moving through a tunnel towards a bright light, or being greeted by deceased loved ones or religious figures.

While these experiences are often associated with feelings of peace and transcendence, some people report encountering frightening or disturbing visions, such as hellish landscapes or demonic figures. The exact nature of these experiences varies widely, with no two NDEs being exactly alike.

Theories of NDEs

Despite the widespread reports of NDEs, there is no scientific consensus on their nature or underlying mechanisms. Many theories have been proposed, including the role of brain chemistry, oxygen deprivation, and the possibility of consciousness existing beyond the physical body.

One popular theory suggests that NDEs may be the result of changes in brain chemistry, such as the release of endorphins or other neurochemicals that produce feelings of euphoria and transcendence. Another theory proposes that NDEs may be a form of oxygen deprivation, causing hallucinations or altered states of consciousness.

Other theories suggest that NDEs may be evidence of consciousness existing beyond the physical body. Some proponents of this theory point to the many cases of people reporting verifiable information during their NDEs, such as events that occurred outside of their field of vision or information about their medical condition that was not known to them at the time.

Applications of NDEs

While the scientific community remains divided on the nature of NDEs, many people who have had these experiences report profound and lasting effects on their lives. Some people report increased spiritual awareness or a renewed appreciation for life, while others report a decrease in anxiety or fear of death.

NDEs have also been used in the field of psychology and psychotherapy, with some therapists using them as a way to help patients deal with issues related to death and dying. However, the use of NDEs in therapy remains controversial, with some critics arguing that they may not be reliable or valid indicators of spiritual or psychological well-being.

The enigma of near-death experiences continues to fascinate scientists, researchers, and the general public alike. While many theories have been proposed to explain the nature and underlying mechanisms of NDEs, no scientific consensus has been reached.

As research in this field continues to progress, we may gain new insights into the nature of NDEs and their potential applications. Until then, the phenomenon of NDEs remains a subject of wonder and mystery, reminding us of the many mysteries of the human mind and consciousness.

65

The Power of Meditation

Meditation is a practice that has been around for thousands of years and is often associated with spiritual and religious traditions. However, in recent years, meditation has gained popularity as a secular practice that can improve mental and physical well-being. Studies have shown that regular meditation can lead to reduced stress, anxiety, and depression, as well as improved cognitive function and overall health. In this article, we will explore the power of meditation and its potential benefits.

- **What is Meditation?**

Meditation is a mental practice that involves focusing one's attention on a specific object, thought, or activity to achieve a state of mental clarity and relaxation. While there are many different types of meditation, most involve some combination of the following elements:

- Focus: directing attention to a specific object or thought.
- Relaxation: releasing tension in the body and mind.
- Breath: using the breath as a focal point to achieve a state of calm and clarity.

Benefits of Meditation

Reduced Stress and Anxiety

Studies have shown that regular meditation can reduce stress and anxiety by decreasing the production of the stress hormone cortisol. This can lead to improved mental and physical health, as well as better overall well-being.

Improved Cognitive Function

Meditation has also been shown to improve cognitive function, including increased focus, concentration, and memory. This is thought to be due to the practice of training the mind to stay focused and present in the moment.

Lower Blood Pressure

Studies have found that regular meditation can lead to lower blood pressure, which can reduce the risk of heart disease and stroke.

Reduced Depression

Meditation has also been found to be an effective treatment for depression, with some studies showing that it can be as effective as medication in treating mild to moderate depression.

Increased Self-Awareness

Meditation can also lead to increased self-awareness, helping individuals to understand their thoughts and emotions more clearly. This can lead to greater self-acceptance and a better understanding of one's own needs and desires.

· How to Start Meditating

Meditation is a practice that can be done anywhere and at any time, making it accessible to anyone. To start meditating, find a quiet space where you can sit comfortably for a few minutes. Start by focusing on your breath, breathing in slowly and deeply through your nose and out through your mouth. If your mind begins to wander, gently redirect your attention back to your breath. You can start with just a few minutes a day and gradually increase the length of your meditation sessions over time.

Meditation is a powerful tool for improving mental and physical well-being, and it has been practiced for thousands of years. With regular practice, meditation can lead to reduced stress and anxiety, improved cognitive function, lower blood pressure, reduced depression, and increased self-awareness. By making meditation a part of your daily routine, you can harness its power to improve your life and overall well-being.

66

The Secret of Lucid Dreaming: Controlling Your Dreams

Lucid dreaming is the ability to recognize and control your dreams while you are in the dream state. This phenomenon has fascinated humans for centuries, as it offers a unique opportunity to explore the depths of the subconscious mind and experience things that are not possible in waking life. In this article, we will explore the secret of lucid dreaming, including its history, techniques, and potential benefits.

The History of Lucid Dreaming

The concept of lucid dreaming can be traced back to ancient cultures, such as the Greeks and Romans, who believed that dreams were a gateway to the divine realm. In the 19th and 20th centuries, the study of lucid dreaming began to gain popularity, with researchers like Frederik van Eeden and Celia Green exploring the phenomenon in depth.

Techniques for Lucid Dreaming

There are many techniques for inducing lucid dreaming, with some methods being more effective than others. Some popular techniques include:

1. Reality testing: This involves regularly asking yourself throughout the day whether you are dreaming or not, to increase your awareness of your state of consciousness.

2. Mnemonic induction of lucid dreams (MILD): This involves setting an intention to remember that you are dreaming and repeating a mantra to yourself as you fall asleep.
3. Wake back to bed (WBTB): This involves waking up after a few hours of sleep and then going back to sleep with the intention of entering a lucid dream.

Benefits of Lucid Dreaming

Increased Creativity

Lucid dreaming has been linked to increased creativity, as it allows individuals to explore their subconscious mind and tap into their innate creativity.

Improved Problem-Solving Skills

Lucid dreaming can also improve problem-solving skills, as it allows individuals to experiment with different solutions in a safe and controlled environment.

Reduced Anxiety

Lucid dreaming can be a useful tool for reducing anxiety, as it can help individuals confront and overcome their fears and phobias in a controlled setting.

Improved Sleep

Lucid dreaming has also been linked to improved sleep quality, as it can help individuals relax and enter a deeper state of rest.

Conclusion

Lucid dreaming is a fascinating phenomenon that has been studied for centuries. With regular practice, anyone can learn to induce lucid dreams and explore the depths of their subconscious mind. Whether you are looking to increase your creativity, improve your problem-solving skills, reduce anxiety, or simply experience something new and exciting, lucid dreaming is a powerful tool that can offer a unique and transformative experience.

67

The Enigma of Déjà Vu: The Feeling of Familiarity

Déjà vu is a common experience that many people have had at some point in their lives. It is the feeling of familiarity or the sense of having lived through a situation before, even though it is new or unfamiliar. Déjà vu is a French term that translates to "already seen," and while it is a common experience, it remains a mystery to scientists and researchers. In this article, we will explore the enigma of déjà vu, including its history, theories, and potential explanations.

The History of Déjà Vu

The concept of déjà vu can be traced back to the ancient Greeks, who believed that it was a sign of divine intervention or prophecy. In the 19th century, French philosopher Émile Boirac coined the term "déjà vu," describing it as a feeling of familiarity with an experience that is happening for the first time. Since then, déjà vu has been the subject of much research and speculation.

Theories of Déjà Vu

There are many theories about the underlying mechanisms of déjà vu, with some suggesting that it is related to memory, perception, or consciousness. Some of the most **popular theories include:**

1. Memory-related: One theory suggests that déjà vu is related to memory, and that it occurs when the brain mistakenly recalls a previous experience as being new. This could be due to similarities between the current experience and a past experience, or due to a glitch in the memory retrieval process.

2. Perception-related: Another theory suggests that déjà vu is related to perception, and that it occurs when the brain misinterprets sensory information, causing a feeling of familiarity. This could be due to similarities in sensory input, or due to a misinterpretation of sensory information.

3. Consciousness-related: A third theory suggests that déjà vu is related to consciousness, and that it occurs when the brain enters a state of heightened awareness or attention. This could be due to changes in brain activity or due to a shift in consciousness.

Potential Explanations for Déjà Vu

While the underlying mechanisms of déjà vu remain unclear, there are several potential explanations for the phenomenon. Some of these include:

1. Neurological Factors: Some researchers believe that déjà vu is related to changes in brain activity, specifically in the hippocampus, which is involved in memory retrieval. It is possible that déjà vu occurs when the brain mistakenly retrieves a memory that is similar to the current experience.

2. Attentional Factors: Another potential explanation for déjà vu is related to attentional factors, specifically in the prefrontal cortex, which is involved in attention and awareness. It is possible that déjà vu occurs when the brain is paying close attention to a situation and misinterprets it as being familiar.

3. Emotional Factors: Some researchers believe that déjà vu may be related to emotions, specifically feelings of familiarity or nostalgia. It is possible that déjà vu occurs when the brain experiences a strong emotional response to a situation and interprets it as being familiar.

Déjà vu is a common experience that has puzzled researchers and scientists for centuries. While there are many theories about its underlying mechanisms, the exact cause of déjà vu remains unknown. However, with ongoing research and study, we may one day unravel the mystery of this enigmatic phenomenon. In the meantime, the experience of déjà vu continues to captivate and intrigue us, reminding us of the many mysteries of the human mind and consciousness.

68

The Power of Brainwaves: Using Neuroscience to Improve Your Life

Over the past few decades, neuroscience has made significant strides in understanding the human brain and its functions. One of the most fascinating areas of research involves brainwaves, the electrical signals generated by neural activity. By harnessing the power of brainwaves, we can potentially improve our cognitive abilities, emotional well-being, and overall quality of life. This article explores the science behind brainwaves, various types of brainwave patterns, and how you can use this knowledge to optimize your mental performance and well-being.

Understanding Brainwaves

Brainwaves are the product of electrical activity in the brain as neurons communicate with each other. These electrical signals can be measured using electroencephalography (EEG) and are classified into different categories based on their frequency:

1. Delta Waves (0.5-4 Hz): These are the slowest brainwaves, typically associated with deep sleep, healing, and regeneration.
2. Theta Waves (4-8 Hz): Theta waves are present during light sleep, meditation, and moments of creativity and insight.
3. Alpha Waves (8-12 Hz): Alpha waves signify a relaxed and calm state

of mind, often experienced during meditation or daydreaming.

4. Beta Waves (12–30 Hz): These waves are associated with active thinking, problem-solving, and decision-making during our waking state.

5. Gamma Waves (30–100 Hz): The fastest brainwaves, gamma waves are linked to higher cognitive functioning, learning, and memory consolidation.

Harnessing the Power of Brainwaves

Understanding these brainwave patterns can help you tailor your daily activities and habits to optimize your mental performance and emotional well-being. Here are some ways to tap into the power of brainwaves:

1. Improve sleep quality: Prioritize sleep hygiene to enhance the quality of your sleep and maximize the benefits of delta and theta waves. Establish a consistent sleep schedule, create a relaxing bedtime routine, and ensure your sleep environment is conducive to rest.

2. Practice mindfulness and meditation: Engaging in regular meditation or mindfulness practices can help increase your alpha waves, promoting relaxation and reducing stress. Techniques such as deep breathing, body scans, or loving-kindness meditation can be highly effective in fostering a calm and focused state of mind.

3. Optimize focus and productivity: To make the most of beta waves, create an environment that minimizes distractions and promotes concentration. Break tasks into manageable chunks, use time-management techniques such as the Pomodoro Technique, and allocate specific periods for focused work.

4. Boost cognitive abilities: Stimulate gamma waves by engaging in activities that challenge your brain, such as puzzles, learning a new language, or playing a musical instrument. These activities can enhance cognitive functions, memory, and learning capabilities.

5. Biofeedback and neurofeedback: These therapies involve monitoring your brainwaves in real-time and using this information to learn how to regulate your brain activity. Biofeedback and neurofeedback can help

improve various aspects of mental health, such as reducing anxiety, managing stress, and enhancing focus.

The power of brainwaves should not be underestimated. By understanding the various brainwave patterns and how they impact our cognitive and emotional states, we can take steps to improve our mental performance and overall well-being.

69

The Mystery of Collective Consciousness: Do We Share a Mind?

The concept of collective consciousness, or the idea that a group of individuals can share a single, unified mind or consciousness, has been a subject of fascination and debate for decades. While some argue that collective consciousness is a real phenomenon, others remain skeptical, citing the lack of scientific evidence. In this article, we will explore the mystery of collective consciousness, including its history, theories, and potential implications.

The History of Collective Consciousness

The concept of collective consciousness can be traced back to the work of French sociologist Émile Durkheim, who used the term to describe the shared beliefs, values, and practices of a particular society or culture. Durkheim believed that collective consciousness was an essential component of social cohesion and integration.

In the 20th century, the concept of collective consciousness began to take on a more mystical and spiritual connotation, with many philosophers and thinkers suggesting that it was evidence of a deeper, more profound interconnectedness between individuals and the universe as a whole.

Theories of Collective Consciousness

Despite the long history of the concept of collective consciousness, there is no scientific consensus on its existence or nature. Many theories have been proposed, including the role of empathy, mirror neurons, and quantum entanglement.

One popular theory suggests that collective consciousness is the result of empathy, or the ability to feel and understand the emotions and experiences of others. According to this theory, when individuals experience intense emotions or thoughts, those emotions and thoughts can be transmitted to others in their vicinity, creating a shared consciousness.

Another theory proposes that collective consciousness is linked to the concept of mirror neurons, which are neurons in the brain that activate when an individual observes someone else performing an action. According to this theory, the activation of mirror neurons creates a shared consciousness that allows individuals to understand and empathize with each other's experiences.

Implications of Collective Consciousness

If collective consciousness is a real phenomenon, it could have profound implications for our understanding of human nature and the universe as a whole. It could suggest that we are all connected in ways that we do not yet fully understand, and that our thoughts and emotions have the power to influence the world around us.

Collective consciousness could also have practical applications, such as in the fields of psychology and healthcare. If we are able to tap into a collective consciousness, it could allow us to better understand and address the needs of others, and to develop more effective treatments for a variety of mental and physical conditions.

The mystery of collective consciousness continues to elude scientists and researchers, with no scientific consensus on its existence or nature. While many theories have been proposed, the evidence remains inconclusive, and more research is needed to fully understand the phenomenon.

Despite this, the concept of collective consciousness remains a subject of fascination and intrigue, reminding us of the many mysteries of the human mind and our interconnectedness with the world around us.

70

The Secrets of Astral Projection: Out-of-Body Experiences

Astral projection, also known as an out-of-body experience (OBE), is a phenomenon that has captivated the human imagination for centuries. It involves the sensation of one's consciousness leaving the physical body and traveling through an alternate plane or dimension, known as the astral plane. This article delves into the secrets of astral projection, exploring its history, reported experiences, scientific perspectives, and practical techniques for inducing an OBE.

· A Glimpse into History

Astral projection has been documented in various cultures and spiritual traditions throughout history. Ancient Egyptian texts describe the "ka" or the soul, which can travel outside the body. In Hinduism, the concept of the "subtle body" or "sukshma sharira" refers to an ethereal body separate from the physical form. The Tibetan Book of the Dead also mentions a similar concept called the "bardo body" that can traverse other realms after death.

· The Astral Projection Experience

People who report experiencing astral projection often describe similar sensations, such as:

A. The Vibrational State: This is a common precursor to astral projection, characterized by a buzzing or vibrating sensation throughout the body.

B. Separation: The individual feels their consciousness detaching from their physical body, which may be accompanied by a floating or rising sensation.

C. The Astral Plane: Once separated, the person's consciousness enters the astral plane, a realm that may appear vivid, dreamlike, or entirely different from the physical world.

D. Reintegration: The experience typically concludes with the individual's consciousness returning to their physical body, often with a jolt or sudden awareness of being "back."

· Scientific Perspectives on Astral Projection

While astral projection has been a topic of interest for centuries, its scientific basis remains debated. Several theories have been proposed to explain the phenomenon:

A. Altered States of Consciousness: Researchers suggest that astral projection might be the result of altered states of consciousness, such as those experienced during deep meditation, lucid dreaming, or near-death experiences.

B. Neurological Explanations: Some scientists argue that OBEs are the result of specific brain activity, such as disruptions in the brain's vestibular system, which is responsible for our sense of balance and spatial orientation.

C. Psychological Explanations: Others propose that astral projection is a form of dissociation or a psychological defense mechanism, allowing individuals to escape from stress, trauma, or other difficult emotions.

· Techniques for Inducing Astral Projection

While there is no foolproof method for inducing astral projection, various techniques have been proposed by practitioners, including:

A. Relaxation and Visualization: Deep relaxation techniques, such as progressive muscle relaxation or focused breathing, can be used in conjunction with visualization exercises to help induce an OBE.

B. The Rope Technique: This popular method involves imagining a rope extending from one's body and mentally "climbing" it to separate the consciousness from the physical form.

C. Binaural Beats: Some people claim that listening to binaural beats, which are audio tracks designed to create specific brainwave patterns, can help facilitate astral projection.

The secrets of astral projection continue to mystify and intrigue us.

VIII

Secrets of Politics and Power

71

The Secrets of the Illuminati

The Illuminati is a secret society that has captured the public's imagination for centuries. Rumors and conspiracy theories surround the group, with many people believing that it has a hand in global events and controls world governments. But what are the secrets of the Illuminati, and is there any truth to the rumors?

Here are some key facts about the Illuminati:

1. The Illuminati was a secret society that existed in the late 18th century. It was founded by a Bavarian philosopher named Adam Weishaupt in 1776.
2. The Illuminati aimed to promote enlightenment and oppose religious and political conservatism. It advocated for individual liberty and freedom of thought.
3. The group's teachings were based on the principles of reason, rationalism, and secularism. It rejected traditional religious beliefs and advocated for a world based on scientific inquiry.
4. The Illuminati was opposed by the Catholic Church and other conservative forces, who saw it as a threat to their power.
5. The Illuminati was disbanded in the late 18th century, but rumors of its continued existence persisted.
6. The Illuminati is often associated with a range of conspiracy theories,

including claims that it controls world governments and the global financial system.

7. The Illuminati is also linked to the New World Order conspiracy theory, which posits that a secretive elite group is plotting to establish a single global government.

8. Some conspiracy theorists believe that the Illuminati is responsible for a range of global events, including the French Revolution and the 9/11 attacks.

9. The Illuminati is often depicted in popular culture as a shadowy and powerful organization, with symbols such as the all-seeing eye and the pyramid featuring prominently.

10. There is little evidence to support the conspiracy theories surrounding the Illuminati. Most scholars believe that the group was simply a small, short-lived organization that had little impact on world events.

While the secrets of the Illuminati remain shrouded in mystery and conspiracy, it is important to separate fact from fiction when exploring this enigmatic group.

72

The Conspiracy Theory of 9/11

The September 11 attacks on the World Trade Center and the Pentagon in 2001 were one of the most significant events in modern history. However, in the aftermath of the attacks, a conspiracy theory emerged that challenged the official narrative of what happened that day. This theory suggests that the attacks were not the work of terrorist group Al-Qaeda, but rather a covert operation orchestrated by the United States government itself.

Here are some key facts about the 9/11 conspiracy theory:

1. The conspiracy theory suggests that the World Trade Center towers were not brought down by the impact of the planes or the resulting fires, but rather by controlled demolition.
2. Proponents of the theory argue that the government used explosives to bring down the buildings, and that the planes were simply a diversion.
3. The conspiracy theory also suggests that the Pentagon was not hit by a commercial airliner, but rather a missile or some other type of military aircraft.
4. Supporters of the theory believe that the U.S. government either knew about the attacks in advance and allowed them to happen, or actively planned and executed them.
5. The 9/11 conspiracy theory has been widely debunked by experts in engineering, physics, and other relevant fields.

6. The theory relies on a number of inconsistencies and inaccuracies in the official narrative of the attacks, as well as the belief that the U.S. government is capable of such a vast and complex conspiracy.
7. Many people who subscribe to the theory are motivated by a distrust of the government and a desire for more transparency and accountability.
8. The 9/11 conspiracy theory has been the subject of numerous books, documentaries, and online communities.
9. The theory has also been used to justify a range of other conspiracy theories, including the belief that the U.S. government was behind the assassination of President John F. Kennedy.
10. Despite widespread debunking, the 9/11 conspiracy theory persists and continues to be a controversial topic in public discourse.

While the events of September 11, 2001 remain a tragic and deeply emotional part of modern history, it is important to approach conspiracy theories with a critical eye and seek out evidence-based explanations. The official narrative of the attacks has been supported by numerous investigations and expert analyses, and while it may not provide all the answers, it remains the most credible account of what happened that day.

73

The Mystery of the JFK Assassination

The assassination of President John F. Kennedy on November 22, 1963, remains one of the most infamous events in American history. Despite numerous investigations and inquiries, the full truth behind the assassination remains shrouded in mystery and controversy. In this article, we will explore some of the key facts and theories surrounding the JFK assassination.

The Official Narrative

According to the official narrative, Lee Harvey Oswald acted alone in shooting President Kennedy from the sixth floor of the Texas School Book Depository in Dallas. Oswald was a former Marine and self-proclaimed Marxist who had defected to the Soviet Union before returning to the United States. He was arrested and charged with the assassination, but was himself shot and killed by nightclub owner Jack Ruby while in police custody just two days later.

The Warren Commission, a government-appointed investigation, concluded in 1964 that Oswald acted alone in shooting Kennedy. However, many people remain skeptical of this conclusion, and numerous alternative theories have emerged over the years.

Conspiracy Theories

One of the most popular conspiracy theories surrounding the JFK assassination is that there was a larger plot involving multiple people and organizations. Some people believe that the CIA, organized crime, or even

Vice President Lyndon B. Johnson were involved in the assassination.

One theory suggests that Oswald was part of a larger conspiracy involving anti-Castro Cuban exiles who were angry with Kennedy for failing to provide support for the Bay of Pigs invasion. Another theory suggests that Kennedy was killed in retaliation for his attempts to dismantle the Federal Reserve and end the Vietnam War.

Another popular theory is that there was a second shooter involved in the assassination. Many people believe that the fatal shot that hit Kennedy in the head came from a different direction than the shots fired by Oswald.

Evidence and Inconsistencies

There are numerous pieces of evidence and inconsistencies that have fueled conspiracy theories about the JFK assassination. For example, the famous Zapruder film, which captured the assassination on film, appears to show Kennedy's head being propelled backward, suggesting that the fatal shot came from the front rather than the rear. Other inconsistencies include the fact that some witnesses reported hearing shots coming from multiple locations, and that the bullet that killed Kennedy was found on a stretcher at the hospital in almost pristine condition.

Another piece of evidence that has fueled conspiracy theories is the so-called "magic bullet" theory. According to this theory, a single bullet fired by Oswald somehow managed to strike both Kennedy and Texas Governor John Connally, who was sitting in front of Kennedy. The Warren Commission concluded that the bullet entered Kennedy's back, exited his throat, then entered Connally's back and exited his chest. However, many people believe that this trajectory is impossible and that it supports the idea of multiple shooters.

The JFK assassination remains one of the most enduring mysteries in American history. Despite numerous investigations and inquiries, the full truth behind the assassination remains elusive. While the official narrative suggests that Oswald acted alone in shooting Kennedy, many people remain skeptical of this conclusion and believe that there was a larger conspiracy involved. While it may be difficult to uncover the truth behind the assassination after all these years, it is important to continue to examine the evidence and consider all possibilities in order to better understand this pivotal moment in American history.

74

The Secret of the Treasurer under Sea

The ocean is a vast and mysterious place, with many secrets waiting to be discovered. One of the most intriguing mysteries of the deep is the existence of sunken treasure ships. For centuries, sailors and pirates have been seeking out these lost treasures, hoping to strike it rich. In this article, we will explore the secret of the treasurer under the sea, the history of sunken treasure, and the modern technology used to uncover these hidden treasures.

The History of Sunken Treasure

Sunken treasure has captured the imaginations of people for centuries. The idea of lost riches hidden deep in the ocean has been the inspiration for countless stories and legends. The reality is that there are many sunken treasure ships lying at the bottom of the ocean, waiting to be discovered.

One of the most famous examples of sunken treasure is the Spanish galleon, the Nuestra Señora de Atocha. The ship was carrying a large quantity of gold, silver, and other precious items when it sank off the coast of Florida in 1622. For centuries, the location of the ship and its treasure was unknown, but in 1985, a salvage team led by treasure hunter Mel Fisher discovered the wreckage and recovered over $400 million worth of treasure.

Another famous example of sunken treasure is the Titanic, which sank in 1912 after hitting an iceberg. While the main focus of the Titanic's legacy is the tragic loss of life, there were also many valuable items on board the ship, including jewelry, artwork, and other personal belongings of the passengers.

Over the years, numerous expeditions have been undertaken to recover these items, although the effort has been met with controversy due to the sensitivity of the site and the potential for damage to the wreckage.

The Technology Used to Discover Sunken Treasure

Over the years, technology has played a crucial role in the discovery and recovery of sunken treasure. In the past, treasure hunters relied on basic tools such as diving gear and metal detectors to search for lost treasure. However, modern technology has enabled treasure hunters to search for sunken treasure more effectively and safely.

One of the most important technological advances in the search for sunken treasure is the use of sonar imaging. Sonar can be used to create detailed images of the ocean floor, allowing treasure hunters to locate potential sites for further investigation. In addition to sonar, underwater robots and submersibles are also used to explore deep sea wrecks and recover valuable items.

Another important tool in the search for sunken treasure is the magne-tometer. This device can detect metal objects buried in the ocean floor, making it easier to locate wrecks that may contain valuable items.

The Legal and Ethical Issues Surrounding Sunken Treasure

While the discovery of sunken treasure can be exciting and lucrative, it is not without legal and ethical issues. The ownership of sunken treasure can be a contentious issue, as many countries have laws governing the ownership of items recovered from their waters. In some cases, sunken treasure may be considered a part of a nation's cultural heritage and cannot be removed from the country.

Another ethical issue surrounding sunken treasure is the potential damage to the underwater environment. Salvage operations can disrupt delicate ecosystems and damage underwater structures. There is also the risk of looting, where valuable artifacts are removed from the wreck and sold illegally.

The search for sunken treasure continues to captivate people around the world. While the discovery of lost treasure can be exciting, it is important to consider the legal and ethical issues surrounding the recovery of sunken treasure. With modern technology and careful consideration, treasure hunters can continue to explore the mysteries of the deep while also preserving the underwater environment and cultural heritage. The secrets of the treasure under the sea will continue to fascinate us for years to come, inspiring adventurers

75

The Enigma of the Freemasons

The Freemasons are a fraternal organization that dates back to the late 16th century. The group has been the subject of much speculation and conspiracy theories, with some claiming that the Freemasons are a secret society with an agenda to control the world.

The origins of the Freemasons can be traced back to the guilds of stonemasons in medieval Europe. These guilds were responsible for building many of the cathedrals and castles of the time, and they had a system of secret signs and symbols to identify themselves to each other.

Over time, the stonemasons' guilds began to admit non-masons into their ranks, and the organization evolved into what is now known as the Freemasons. Today, the Freemasons are a fraternal organization that promotes personal development and charitable work.

However, the Freemasons have also been the subject of much speculation and conspiracy theories. Some claim that the Freemasons are a secret society with an agenda to control the world. Others allege that the Freemasons have been involved in historical events such as the French Revolution and the assassination of U.S. President John F. Kennedy.

The Freemasons themselves maintain that they are simply a fraternal organization that promotes personal development and charitable work. However, the secretive nature of the organization has fueled conspiracy theories and speculation for centuries.

One of the most well-known symbols of the Freemasons is the square and compass. This symbol represents the organization's commitment to morality and the pursuit of truth. The symbol is also associated with the idea of "squaring the circle," which is a reference to the ancient Greek problem of constructing a square with the same area as a given circle.

The Freemasons also have a system of secret signs and symbols that are used to identify members to each other. This system of secret signs and symbols has led to much speculation about the organization's true motives and intentions.

One of the most popular conspiracy theories about the Freemasons is that they are a secret society with an agenda to control the world. According to this theory, the Freemasons are working behind the scenes to influence politics and the media in order to promote their own interests.

Another conspiracy theory about the Freemasons is that they were involved in the French Revolution. According to this theory, the Freemasons used their influence to foment the revolution and overthrow the French monarchy.

There are also many conspiracy theories about the Freemasons' involvement in the assassination of U.S. President John F. Kennedy. Some claim that Lee Harvey Oswald, the man who was accused of assassinating Kennedy, was a Freemason and that the organization was involved in a conspiracy to kill the president.

Despite these conspiracy theories, the Freemasons maintain that they are simply a fraternal organization that promotes personal development and charitable work. The organization has been involved in many charitable projects over the years, including supporting hospitals and schools, and providing disaster relief.

In recent years, the Freemasons have become more open about their organization, and many lodges now welcome visitors and provide information about the organization's history and goals. However, the secretive nature of the organization has led to continued speculation and conspiracy theories.

In conclusion, the Freemasons are a fraternal organization that dates back to the late 16th century. While the organization promotes personal development and charitable work, it has also been the subject of much speculation and conspiracy theories over the years. Despite this speculation, the Freemasons maintain that they are simply a fraternal organization that promotes personal development and charitable work.

76

The Power of Propaganda: How Governments Manipulate the Masses

Propaganda is a powerful tool that governments and organizations have used for centuries to shape public opinion and influence behavior. It is a form of communication that is designed to influence people's beliefs, attitudes, and actions towards a particular cause or idea. The use of propaganda has been evident in some of the most significant events in history, such as World War II and the Cold War. In this article, we will explore the power of propaganda and how it has been used throughout history to manipulate the masses.

The History of Propaganda

The origins of propaganda can be traced back to ancient times when rulers used symbols and images to communicate with their subjects. However, the term "propaganda" was first used in the 17th century by the Catholic Church to describe its efforts to promote Catholicism and counter the Protestant Reformation.

During World War I and World War II, propaganda played a significant role in shaping public opinion and rallying support for the war effort. Governments used propaganda to demonize the enemy, promote patriotism, and recruit soldiers. Propaganda posters, films, and radio broadcasts were widely used to convey these messages to the public.

In the years following World War II, the Cold War between the United

States and the Soviet Union led to an increase in propaganda on both sides. The United States government used propaganda to promote democracy and capitalism, while the Soviet Union used propaganda to promote communism and socialism.

The Power of Propaganda

Propaganda is a powerful tool because it appeals to people's emotions and can bypass critical thinking. By presenting a message in a way that is appealing and easy to understand, propaganda can influence people's beliefs and attitudes towards a particular issue or idea.

One of the ways in which propaganda is powerful is through its ability to create a sense of "us vs. them." By demonizing an enemy or promoting a particular ideology, propaganda can create a sense of unity among a group of people. This can be particularly effective during times of war or political conflict.

Another way in which propaganda is powerful is through its ability to shape public opinion. Through the use of carefully crafted messages and images, propaganda can influence how people view a particular issue or idea. This can be particularly effective when the propaganda is repeated over time, as people may begin to accept the message as true.

Examples of Propaganda

There have been many examples throughout history of propaganda being used to manipulate the masses. One of the most well-known examples is Nazi propaganda during World War II. The Nazi regime used propaganda to promote the idea of Aryan supremacy and demonize Jews and other groups deemed "undesirable." This propaganda played a significant role in the Holocaust and other atrocities committed by the Nazi regime.

During the Cold War, both the United States and Soviet Union used propaganda to promote their ideologies and demonize the other side. The United States used propaganda to promote democracy and capitalism, while the Soviet Union used propaganda to promote communism and socialism.

In recent years, social media has become a powerful tool for propaganda. In countries like Russia and China, governments use social media to shape public opinion and promote their political agendas. This can be particularly effective because social media allows propaganda to be shared quickly and widely, often without fact-checking or critical analysis.

The Ethics of Propaganda

The use of propaganda raises ethical questions about the manipulation of the masses. While propaganda can be used for positive purposes, such as promoting public health or safety, it can also be used for nefarious purposes, such as promoting hate or inciting violence.

There is also the question of who should be responsible for regulating propaganda. In some countries, the government is responsible for regulating propaganda, while in others, it is left up to the media or civil society organizations.

77

The Secret of the Deep State: The Hidden Influence of Government Agencies

The concept of the "Deep State" refers to a network of powerful individuals and institutions within a government that operate behind the scenes to influence policy and decision-making. These individuals and institutions may include intelligence agencies, military contractors, lobbyists, and other government bureaucrats.

The idea of a Deep State is often associated with conspiracy theories, but there is some evidence to suggest that there may be some truth to it. Some argue that the influence of these individuals and institutions can be seen in government decisions that do not seem to align with the interests of the general public.

One example of the Deep State at work is the influence of the military-industrial complex. This network of defense contractors, military officials, and government bureaucrats has a vested interest in maintaining a state of perpetual war, as it provides them with lucrative contracts and job security.

Another example is the influence of intelligence agencies such as the CIA and the NSA. These agencies have been accused of engaging in illegal activities such as domestic spying, torture, and assassination, all in the name of national security.

Some argue that the influence of the Deep State can be seen in the reluctance of politicians to pursue policies that are in the best interests of the general public. Instead, politicians may prioritize the interests of the Deep State in order to maintain their own power and influence.

The Deep State may also be involved in influencing the media, with some suggesting that major news outlets are controlled by a small group of powerful individuals who use the media to shape public opinion and promote their own interests.

Despite the controversy surrounding the concept of the Deep State, there is some evidence to suggest that there may be some truth to it. However, it is important to be skeptical of conspiracy theories and to seek out reliable sources of information when evaluating claims about the influence of the Deep State.

It is also important to recognize that the influence of the Deep State is not necessarily a sign of a nefarious conspiracy. In many cases, the individuals and institutions that make up the Deep State may simply be pursuing their own interests in a way that is not aligned with the interests of the general public.

To better understand the role of the Deep State in government decision-making, it is important to remain vigilant and informed about the actions of government officials and institutions. This may involve researching government policies and decisions, seeking out alternative sources of information, and participating in grassroots activism and political organizing.

The concept of the Deep State refers to a network of powerful individuals and institutions within a government that operate behind the scenes to influence policy and decision-making. While there is some evidence to suggest that the Deep State may be real, it is important to be skeptical of conspiracy theories and to seek out reliable sources of information when evaluating claims about the influence of the Deep State. Ultimately, it is up to individuals to remain vigilant and informed about the actions of government officials and institutions, in order to ensure that government policies and decisions are in the best interests of the general public.

78

The Enigma of Political Assassinations: The Deaths of MLK, RFK, and more

Political assassinations have long been a topic of fascination and speculation, with the deaths of prominent figures such as Martin Luther King Jr. and Robert F. Kennedy still shrouded in mystery and controversy. These assassinations have left lasting scars on the political landscape of the United States, and have fueled conspiracy theories and speculation about the true motives behind these killings.

One of the most famous political assassinations in American history is the assassination of President John F. Kennedy. While Lee Harvey Oswald was officially charged with the murder, conspiracy theories have swirled around the case for decades. Some believe that the CIA was involved, while others point to organized crime or even Cuban exiles seeking revenge for the failed Bay of Pigs invasion.

The assassination of Martin Luther King Jr. is another case that has generated controversy and speculation. While James Earl Ray was convicted of the killing, many believe that he was not acting alone. Some have pointed to the involvement of the FBI or other government agencies, citing evidence of wiretapping and surveillance of King.

The assassination of Robert F. Kennedy, brother of President John F. Kennedy, is yet another case that has generated controversy. Sirhan Sirhan was convicted of the killing, but many have questioned whether he was acting alone. Some have pointed to the involvement of organized crime or even the CIA, citing evidence of suspicious behavior by security personnel at the scene of the shooting.

Despite the many conspiracy theories and controversies surrounding political assassinations, it is important to remember that these tragedies represent a very real loss of life and a serious threat to the democratic process. Assassinations can have a chilling effect on political discourse and can undermine the faith of the public in their elected officials and institutions.

In many cases, political assassinations are the result of complex social and political dynamics that are difficult to unravel. These dynamics may include tensions between different factions within a political movement, conflicts between different branches of government, or deep-seated social and economic inequalities that create a sense of desperation and hopelessness.

Ultimately, the best way to prevent political assassinations is to address the underlying social and political dynamics that can lead to violence and extremism. This may involve addressing issues such as poverty, inequality, and political polarization, as well as investing in education and other forms of social and cultural development.

It is also important to hold those responsible for political assassinations accountable for their actions. This may involve pursuing criminal charges against those who are involved in the killings, as well as supporting efforts to uncover the truth behind these tragic events.

The enigma of political assassinations continues to fascinate and intrigue people around the world. While the true motives behind these killings may never be fully understood, it is important to remember that they represent a serious threat to the democratic process and a very real loss of life. By addressing the underlying social and political dynamics that can lead to violence and extremism, and by holding those responsible for political assassinations accountable for their actions, we can work to prevent these tragedies from happening in the future.

79

The Mystery of Secret Societies: From the Illuminati to Skull and Bones

Throughout history, secret societies have captured the public imagination, fueling conspiracy theories and legends about their true purposes and powers. From the Illuminati to Skull and Bones, these organizations have been the subject of countless books, films, and television shows, and have been the source of speculation and intrigue for generations.

One of the most famous secret societies in history is the Illuminati. Founded in Bavaria in the late 18th century, the Illuminati was a secret society that aimed to promote enlightenment ideals and challenge the power of the Catholic Church and other traditional institutions. The group was believed to have been disbanded by the government, but conspiracy theories about the Illuminati have persisted to this day.

Another famous secret society is Skull and Bones, an elite fraternity at Yale University that has produced many prominent politicians and business leaders. The society's rituals and activities are shrouded in secrecy, leading to speculation about its true purposes and influence on American politics.

Other secret societies that have captured the public imagination include the Freemasons, the Bilderberg Group, and the Knights Templar. These organizations have been the subject of countless conspiracy theories and legends, with some people believing that they are involved in everything from world domination to alien conspiracies.

Despite the many conspiracy theories and legends surrounding secret societies, it is important to remember that not all of these organizations are necessarily sinister or nefarious. Many secret societies, such as the Freemasons, are simply fraternal organizations that provide a sense of community and shared values for their members.

However, it is also true that some secret societies have been involved in criminal or unethical activities. The Sicilian Mafia, for example, is often described as a secret society that uses violence and intimidation to maintain power and control in certain regions of Italy.

Regardless of their true purposes and activities, secret societies continue to capture the public imagination and fuel conspiracy theories and legends. Whether they are promoting enlightenment ideals or engaged in criminal activities, these organizations remain shrouded in mystery and intrigue.

So why do secret societies continue to fascinate us? Perhaps it is because they represent a world beyond the everyday, a realm of hidden knowledge and power that is only accessible to a select few. Perhaps it is because they challenge our assumptions about the way the world works, forcing us to question the motives and actions of those in power.

Whatever the reason, it is clear that secret societies will continue to be a subject of fascination and intrigue for generations to come. Whether they are real or imagined, they represent a glimpse into a world beyond the ordinary, a world of mystery and secrets that will always capture the human imagination.

The mystery of secret societies is a topic that has fascinated people for centuries. From the Illuminati to Skull and Bones, these organizations have captured the public imagination and fueled countless conspiracy theories and legends. While some secret societies are simply fraternal organizations that provide a sense of community and shared values for their members, others have been involved in criminal or unethical activities. Regardless of their true purposes and activities, secret societies represent a world beyond the everyday, a realm of hidden knowledge and power that will always capture the human imagination.

80

The Secrets of the New World Order: A Global Conspiracy or Just a Theory?

The New World Order is a term that has been used for decades to describe a supposed global conspiracy that seeks to establish a single, totalitarian government that controls every aspect of human life. This theory posits that a secret group of elites, including politicians, business leaders, and others, are working together to bring about this new world order and achieve their own agenda.

The origins of the New World Order theory can be traced back to the late 19th century, when the idea of a global government first began to gain traction. However, it was not until the 20th century that the theory really took hold, fueled by events such as World War I and II, the Cold War, and the rise of globalization.

One of the key claims of the New World Order theory is that a secretive group of elites, sometimes referred to as the "globalists," are working together to establish a one-world government that will control every aspect of human life. This government is said to be controlled by a small group of people who hold immense wealth and power, and who are using their influence to manipulate political systems and economies around the world.

Proponents of the theory point to various pieces of evidence to support their claims, including the existence of secretive organizations like the

Bilderberg Group and the Council on Foreign Relations, which are said to be behind the scenes orchestrating events to further their own agenda. They also point to the influence of multinational corporations, which are said to be working hand-in-hand with governments to create a global economy that benefits only a select few.

Opponents of the New World Order theory argue that there is little evidence to support its claims, and that it is simply a conspiracy theory with no basis in reality. They point out that the idea of a global government controlling every aspect of human life is highly unlikely, and that there are too many competing interests and power centers in the world for such a government to exist.

Despite the controversy surrounding the New World Order theory, it has continued to captivate the public imagination and remains a topic of discussion in political and conspiracy circles. Some people believe that it is a very real threat to freedom and democracy, while others see it as nothing more than a fanciful conspiracy theory with no basis in fact.

So, what is the truth about the New World Order? While it is difficult to say for certain, it seems unlikely that a single, all-powerful government could ever be established that would control every aspect of human life. However, it is certainly possible that there are powerful groups and individuals who are working behind the scenes to further their own interests, and that these interests may not always align with the best interests of the general public.

In the end, the truth about the New World Order may never be fully known. But regardless of whether it is a real threat or simply a conspiracy theory, it is important for people to stay informed and engaged in the political process, and to work towards creating a more just and equitable world for everyone. Only by remaining vigilant and active can we ensure that our democratic freedoms and values are protected for generations to come.

IX

Secrets of the Future

81

Colonizing the Solar System

Colonizing the Solar System: Humanity's Next Frontier

As humanity looks to the stars, the prospect of colonizing the solar system looms closer than ever. In the grand scheme of cosmic exploration, humanity has only just begun to take its first steps. With ambitious goals and visionary projects, humankind is on the cusp of a new age of interplanetary colonization. From Mars to Europa, the possibilities are endless as we move further into the cosmos. In this article, we will delve into the challenges and opportunities that await as we set our sights on colonizing the solar system.

The Need for Colonization

The Earth's finite resources and growing population have led to an increasing demand for resources and living space. As a result, scientists and engineers are pushing the boundaries of space exploration, seeking new opportunities for humanity to expand its reach. Colonizing the solar system not only offers potential solutions to these issues but also provides a chance to ensure the long-term survival of our species, should Earth face catastrophic events.

Mars: The Red Planet Beckons

Mars has been a prime target for colonization for decades. Its relatively close proximity to Earth, along with its similarities in size and composition, make it an ideal candidate for human habitation. Currently, several organizations, including SpaceX, are actively working on the technology necessary to establish a permanent human presence on Mars. Challenges include developing sustainable habitats, producing food, and ensuring adequate supplies of water and oxygen. However, the rapid advancements in technology, such as the development of reusable rockets, have made these goals increasingly achievable.

The Moon: A Gateway to the Solar System

The Moon, as our closest celestial neighbor, presents an essential stepping stone in our journey to colonize the solar system. Its potential as a base for further exploration and resource extraction is immense. In recent years, nations and private companies have renewed their interest in lunar missions, with plans for the establishment of lunar bases and mining operations. The Moon's low gravity and proximity to Earth make it an ideal location for launching missions to more distant targets, such as Mars and beyond.

Asteroids: The Cosmic Goldmines

Asteroids, remnants from the early solar system, are abundant in valuable minerals and resources. The potential for asteroid mining is enormous, as these celestial bodies contain metals like platinum, gold, and iron, as well as water and other essential elements. Companies like Planetary Resources and Deep Space Industries are developing the necessary technology for asteroid mining, which could provide critical resources for future space missions and settlements.

Outer Solar System: Europa and Titan

The outer solar system also holds promise for colonization. Jupiter's moon Europa and Saturn's moon Titan are two of the most intriguing candidates. Europa is believed to have a subsurface ocean beneath its icy crust, which could potentially harbor life. Titan, on the other hand, has a thick atmosphere and liquid hydrocarbon lakes, providing opportunities for scientific research and the development of new technologies.

Challenges and Opportunities

Colonizing the solar system comes with numerous challenges, from creating sustainable living environments to the development of advanced propulsion systems. Overcoming these obstacles will require international collaboration, investment in research and technology, and a long-term vision for humanity's future.

However, the opportunities presented by space colonization are vast. Not only can we unlock the potential for resource extraction, scientific discovery, and technological development, but we also have a chance to ensure humanity's survival in the face of global challenges.

As we embark on this epic journey to colonize the solar system, we must remember the immense responsibility that comes with such an endeavor. The future of humanity depends on our ability to work together, innovate, and adapt to the harsh conditions of space. If we rise to the challenge, the solar system will undoubtedly open up a new

82

Mining the Asteroid Belt

The prospect of mining the asteroid belt is an exciting and revolutionary concept that has the potential to redefine space exploration and resource management. Located between the orbits of Mars and Jupiter, the asteroid belt is a vast region of space filled with countless celestial bodies, rich in valuable materials. As humanity continues to push the boundaries of space exploration, the prospect of tapping into this wealth of resources becomes increasingly plausible. This article explores the potential for asteroid mining, the challenges that must be overcome, and the potential impact of this endeavor on our future.

A Treasure Trove in Space

The asteroid belt is comprised of millions of celestial bodies, ranging in size from small rocks to objects hundreds of kilometers in diameter. These asteroids are remnants from the early days of our solar system and are composed of a wide variety of materials. Notably, they contain significant quantities of precious metals, such as platinum, gold, and silver, as well as industrial metals like iron, nickel, and cobalt. Additionally, asteroids can harbor water and other volatile compounds, which can be utilized for life support and fuel production in space.

Given the finite resources on Earth and the growing demand for raw materials, the asteroid belt presents an opportunity to access an almost limitless supply of valuable resources. The economic implications of this are staggering, with some estimates valuing the asteroid belt's resources at hundreds of trillions of dollars.

Current Progress and Technology

The concept of asteroid mining has rapidly transitioned from science fiction to reality, thanks to advancements in technology and increased interest from both the public and private sectors. Several companies, such as Planetary Resources and Deep Space Industries, are actively developing the technology necessary to extract and process materials from asteroids.

The process of asteroid mining begins with the identification and tracking of suitable targets. Through the use of telescopes and advanced imaging techniques, scientists can determine the composition and trajectory of nearby asteroids. Once a suitable candidate has been identified, the next step is to intercept the asteroid and land on its surface.

Several methods are being explored for the extraction of materials from asteroids. One approach involves the use of robotic excavators and drills to break up the surface and collect samples. Another method is to use concentrated sunlight or heating elements to vaporize surface materials, which can then be captured and processed.

Once the materials have been extracted, they must be processed and refined for use. This can be done either on-site using compact, automated systems or by returning the raw materials to Earth or a nearby processing facility, such as a lunar base.

Challenges and Solutions

Mining the asteroid belt presents numerous challenges, from the technical aspects of space travel and resource extraction to the economic and legal frameworks required to support such operations. Some of the key challenges include:

1. Space travel: Traveling to the asteroid belt requires advanced propulsion systems and efficient spacecraft designs. Current propulsion technologies, such as chemical rockets, are not ideal for long-duration missions. However, innovations in electric propulsion and nuclear thermal propulsion offer potential solutions for efficient space travel.

2. Microgravity: The low-gravity environment of asteroids presents unique challenges for mining operations. Anchoring equipment and managing the movement of materials in microgravity require innovative engineering solutions. Potential solutions include the use of harpoons, clamps, or adhesive materials to secure equipment to the asteroid's surface.

3. Communication and control: The vast distances between the asteroid belt and Earth make real-time communication and control of mining equipment difficult. To address this challenge, advanced autonomous systems and artificial intelligence will be required to perform complex tasks with minimal human intervention.

Legal and regulatory framework: The legal landscape for asteroid mining is still under development. The Outer Space Treaty of 1967, which provides the foundation for international space law, does not directly address the issue of asteroid mining. As a result, there is a need to establish a clear regulatory framework to govern the extraction and ownership of resources from celestial bodies. Efforts are underway, such as the adoption of the U.S. Commercial Space Launch Competitiveness Act in 2015, which grants U.S. citizens the right to own resources extracted from celestial bodies.

5.Economic feasibility: Asteroid mining operations require significant initial investments in research, development, and infrastructure. Ensuring the long-term profitability of these ventures is crucial for their success. As technology advances and the costs of space travel decrease, the economic viability of asteroid mining is expected to improve.

6.Environmental and social concerns: Mining operations, whether on Earth or in space, often raise concerns about their impact on the environment and local communities. It is essential to develop sustainable and responsible practices for asteroid mining, minimizing waste and ensuring that the benefits are equitably distributed.

The Future of Asteroid Mining

Despite the challenges associated with asteroid mining, the potential benefits are immense. The successful extraction and utilization of asteroid resources could lead to a new era of space exploration, industrialization, and economic growth. The development of a space-based economy could alleviate pressure on Earth's limited resources and provide opportunities for scientific research, technological advancement, and even the establishment of human settlements beyond our planet.

In the coming decades, asteroid mining could play a critical role in the expansion of humanity's presence in the solar system. As we continue to explore and develop the necessary technologies, the dream of tapping into the vast resources of the asteroid belt becomes increasingly attainable. The future of asteroid mining promises to reshape our understanding of space exploration and our place in the cosmos, opening up a new frontier of possibilities for human ingenuity and ambition.

1.

83

The Search for Extraterrestrial Life

The search for extraterrestrial life has been a topic of fascination and speculation for centuries. As our understanding of the universe has grown, so too has our ability to search for signs of life beyond Earth. Over the past few decades, scientists have made significant progress in this area, using a variety of techniques to scan the cosmos for signs of alien life. In this article, we will explore the history of the search for extraterrestrial life and the methods scientists are using today to uncover its existence.

The History of the Search for Extraterrestrial Life

The idea that there might be other forms of life in the universe has been around for thousands of years. Ancient Greek philosophers such as Democritus and Epicurus believed that there were other worlds like our own, with their own inhabitants. However, it wasn't until the 19th century that scientists began to take the idea of extraterrestrial life seriously.

In 1859, Charles Darwin published his famous book "On the Origin of Species," which introduced the idea of evolution by natural selection. This theory suggested that life on Earth had evolved over millions of years from simple organisms to the complex array of species we see today. If life on Earth could evolve, why not on other planets?

The idea of life on other planets gained further traction in the early 20th century with the discovery of the first exoplanet, a planet orbiting a star other than our sun. As astronomers continued to discover more exoplanets, the search for extraterrestrial life became more and more focused.

The 1960s and 70s were a particularly exciting time for the search for extraterrestrial life. In 1961, astronomer Frank Drake developed the Drake Equation, a formula that estimates the number of intelligent civilizations in our galaxy that we might be able to communicate with. The equation takes into account factors such as the number of stars in the galaxy, the number of habitable planets, and the likelihood of life evolving on those planets.

In 1974, the Arecibo Message was sent from the Arecibo Observatory in Puerto Rico. This was a message aimed at any extraterrestrial civilizations that might be listening, describing our planet and its inhabitants in binary code. While it's highly unlikely that the message will ever be received or understood, it was an important milestone in our search for extraterrestrial life.

The Search for Extraterrestrial Life Today

Today, scientists are using a variety of techniques to search for signs of extraterrestrial life. One of the most promising methods is the search for biosignatures, which are signs of life that can be detected remotely. These might include the presence of certain gases in a planet's atmosphere, such as oxygen or methane, which are produced by living organisms.

The search for biosignatures has become more sophisticated in recent years thanks to advances in technology. For example, the James Webb Space Telescope, set to launch in 2021, will be able to analyze the atmospheres of exoplanets in greater detail than ever before. This could allow scientists to detect the presence of biosignatures and potentially confirm the existence of extraterrestrial life.

Another method being used to search for extraterrestrial life is the study of extremophiles, organisms that can survive in extreme conditions on Earth. By studying these organisms, scientists hope to gain a better understanding of the types of environments that might support life elsewhere in the universe. For example, the discovery of bacteria living in extreme environments such as deep-sea vents and polar ice caps has led scientists to speculate about the possibility of life on other icy worlds in our solar system, such as Europa and Enceladus.

In addition to these methods, scientists are also searching for direct evidence of extraterrestrial life. One of the most ambitious projects in this area is the Search for Extraterrestrial Intelligence (SETI), which uses radio telescopes to listen for signals from other civilizations. While SETI has been scanning the skies for decades, so far there has been no conclusive evidence of intelligent life beyond Earth.

Despite the lack of direct evidence, the search for extraterrestrial life continues to be a major area of scientific research. In recent years, there has been a renewed interest in this field, fueled in part by the discovery of potentially habitable exoplanets in our galactic neighborhood. For example, in 2017, astronomers discovered a system of seven Earth-sized planets orbiting a star called TRAPPIST-1, three of which are in the star's habitable zone.

The discovery of these exoplanets has sparked renewed efforts to search for signs of life beyond Earth. In the coming years, new telescopes and instruments will be launched that will allow scientists to explore the atmospheres of these planets in greater detail. It's possible that we could detect the first biosignatures within the next decade, which would be a major breakthrough in the search for extraterrestrial life.

Implications of Finding Extraterrestrial Life

The discovery of extraterrestrial life would have profound implications for our understanding of the universe and our place within it. It would also raise a host of philosophical and ethical questions, such as whether we have a responsibility to protect other forms of life in the universe.

If we were to discover intelligent life, it could have even more profound implications. We would need to consider how we would communicate with these beings and what kind of ethical and moral considerations we would need to take into account. It's possible that we could learn valuable lessons from these civilizations, or even develop new technologies based on their knowledge.

The search for extraterrestrial life has come a long way since the ancient Greeks first speculated about the existence of other worlds. Thanks to advances in technology and our growing understanding of the universe, we are closer than ever to discovering the first signs of life beyond Earth. While the search is still ongoing, the potential implications of finding extraterrestrial life are truly profound, and could change our understanding of the universe and our place within it forever.

84

The Future of Space Tourism

Space tourism has long been a dream for many people around the world. The idea of being able to travel beyond our planet and experience the wonders of space is incredibly alluring. In recent years, space tourism has become more of a reality, with a number of companies working on developing spacecraft and other technologies that will make it possible for private citizens to travel to space. In this article, we will explore the future of space tourism and what it might look like in the years to come.

The Early Days of Space Tourism

The idea of space tourism has been around for decades, but it wasn't until the early 2000s that it began to move from science fiction to reality. In 2001, businessman Dennis Tito became the first private citizen to travel to space, paying $20 million for a trip to the International Space Station aboard a Russian Soyuz spacecraft.

This was followed by a handful of other space tourists, including Mark Shuttleworth and Anousheh Ansari, who traveled to the International Space Station in 2002 and 2006, respectively. However, these early space tourists were all trained astronauts, and the trips were organized and operated by the Russian space agency.

Commercial Space Tourism Today

Today, several private companies are working to make space tourism more accessible to the general public. The most well-known of these companies is SpaceX, founded by entrepreneur Elon Musk. In addition to its work on developing spacecraft for NASA, SpaceX is also working on developing a spacecraft for space tourism called Dragon.

Other companies in the space tourism industry include Blue Origin, founded by Amazon CEO Jeff Bezos, and Virgin Galactic, founded by Richard Branson. Both companies are working on developing spacecraft that can take private citizens on suborbital flights, allowing them to experience weightlessness and see the curvature of the Earth.

While the cost of space tourism is still prohibitively expensive for most people, it is expected to become more affordable as the technology develops and competition between companies heats up. The cost of a suborbital flight on Virgin Galactic, for example, is currently $250,000, but the company has said that it hopes to eventually bring the cost down to $40,000.

The Future of Space Tourism

As technology continues to develop, the future of space tourism looks increasingly exciting. Here are some of the key trends that are likely to shape the industry in the years to come.

1.Longer Trips

Currently, most space tourism trips are relatively short, lasting just a few hours or days. However, as technology improves and spacecraft become more advanced, it's likely that we will see longer trips become possible. For example, SpaceX is currently working on developing a spacecraft that can take humans to Mars, and it's possible that we could see the first tourist trips to the red planet within the next decade.

2.Space Hotels

Another trend that is likely to emerge in the coming years is the development of space hotels. Several companies, including Orbital Assembly Corporation and Axiom Space, are working on developing space hotels that could host tourists for longer stays in orbit. These hotels would likely offer amenities such as restaurants, gyms, and even space walks.

Space Tourism Beyond Earth Orbit

While most space tourism trips today are focused on suborbital flights or trips to the International Space Station, in the future we could see tourism ventures beyond Earth's orbit. For example, SpaceX has announced plans to send a private citizen on a trip around the moon as early as 2023, and other companies are working on developing spacecraft that could take tourists to asteroids or even Mars.

Advances in Technology

As technology continues to improve, it's likely that we will see new and innovative ways to experience space tourism. For example, companies could develop virtual reality experiences that allow people to experience the sensation of weightlessness without leaving Earth, or create new types of spacecraft that offer more immersive and exciting experiences for tourists.

Environmental Impact

As space tourism becomes more popular, it's important to consider the environmental impact of these trips. The emissions from spacecraft and rockets can have a significant impact on the atmosphere, and it's important for companies to develop sustainable and eco-friendly technologies as they work to make space tourism more accessible.

Safety Concerns

Safety is always a concern when it comes to space travel, and this is particularly true for space tourism. While companies are working hard to develop safe and reliable spacecraft, there is always a risk involved when traveling to space. It's important for companies to prioritize safety and ensure that tourists are fully prepared and informed before embarking on any space tourism trips.

The future of space tourism looks incredibly exciting, with longer trips, space hotels, and trips beyond Earth's orbit all on the horizon. While the industry is still in its early stages, it's clear that there is a growing demand for these types of experiences, and as technology continues to improve, it's likely that space tourism will become more accessible and affordable for the general public.

However, it's important to consider the environmental impact and safety concerns associated with space tourism. Companies must prioritize sustainability and safety in their development of new spacecraft and technologies, and ensure that tourists are fully informed and prepared for the risks involved in traveling to space.

Overall, space tourism represents a new frontier in the travel industry, offering a once-in-a-lifetime experience that is truly out of this world. While it may be some time before space tourism becomes as commonplace as air travel, it's clear that the industry is on the cusp of a new era, with limitless possibilities for those who are bold enough to venture beyond our planet's atmosphere.

85

The Evolution of Space Technology

The evolution of space technology has been a remarkable journey, marked by incredible achievements, innovation, and the indomitable human spirit. From the early days of rocketry to the development of reusable spacecraft and advanced propulsion systems, our understanding and capabilities in space exploration have grown exponentially. In this article, we will take a deep dive into the history of space technology and explore the key milestones that have shaped our understanding of the cosmos and our place within it.

The Dawn of Rocketry

The origins of space technology can be traced back to the early 20th century, with the pioneering work of visionaries such as Konstantin Tsiolkovsky, Robert Goddard, and Hermann Oberth. These early rocket scientists laid the groundwork for modern rocketry, developing the fundamental principles and equations that govern the motion of rockets and their propulsion systems.

In the 1940s and 1950s, the development of rocket technology accelerated with the advent of the V-2 rocket, designed by Wernher von Braun during World War II. After the war, both the United States and the Soviet Union recognized the potential of rockets for military and scientific applications and began to invest heavily in their development.

The Space Race and the Birth of the Space Age

The Space Race between the United States and the Soviet Union marked a turning point in the evolution of space technology. This intense period of competition, which lasted from the late 1950s to the early 1970s, saw a rapid acceleration of advancements in rocketry and space exploration.

In 1957, the Soviet Union launched Sputnik 1, the world's first artificial satellite, ushering in the dawn of the Space Age. This accomplishment was soon followed by a series of other firsts: Yuri Gagarin became the first human to travel to space in 1961, and Valentina Tereshkova became the first woman to do so in 1963.

The United States, spurred on by the Soviet Union's achievements, set its sights on landing a man on the Moon. The Apollo program, led by NASA, was born out of this ambition, and in 1969, Apollo 11 astronauts Neil Armstrong and Buzz Aldrin became the first humans to set foot on the lunar surface.

Satellites and Space Probes: Expanding our Reach

The development and deployment of satellites have played a crucial role in the evolution of space technology. Satellites have revolutionized global communications, weather forecasting, and navigation, among other fields. The launch of the first communication satellite, Telstar 1, in 1962 marked a significant milestone in this area.

Space probes have also allowed us to explore our solar system and beyond. Pioneering missions, such as the Voyager program, have provided us with invaluable information about the planets, moons, and other celestial bodies in our cosmic neighborhood. Launched in 1977, Voyager 1 and 2 have now reached interstellar space, continuing to send back data as they journey further into the unknown.

The Space Shuttle Era

In the 1970s, NASA embarked on an ambitious project to develop a reusable spacecraft that could revolutionize space travel. The Space Shuttle, introduced in 1981, was the world's first reusable spacecraft, capable of carrying astronauts and payloads to and from low Earth orbit.

Over the course of its 30-year operational history, the Space Shuttle program facilitated numerous groundbreaking missions, including the deployment of the Hubble Space Telescope, the construction of the International Space Station (ISS), and various scientific experiments and satellite launches.

However, the Space Shuttle program was not without its challenges. The tragic loss of the Challenger in 1986 and the Columbia in 2003 underscored the risks associated with human spaceflight and the need for continued innovation in safety and technology.

The International Space Station: A Testament to Global Collaboration

The International Space Station (ISS), a collaborative effort between the United States, Russia, Europe, Canada, and Japan, has been a cornerstone of human space exploration since its first module was launched in 1998. The ISS serves as a laboratory, observatory, and living space for astronauts, enabling long-duration missions and groundbreaking scientific research.

The construction and ongoing operation of the ISS have not only expanded our understanding of the challenges and possibilities of living in space but also demonstrated the power of international collaboration in the pursuit of common goals.

The Rise of Private Space Companies

The 21st century has seen a surge in the development of private space companies, such as SpaceX, Blue Origin, and Virgin Galactic. These companies are challenging the traditional model of space exploration, bringing innovation and competition to the field.

SpaceX, founded by Elon Musk in 2002, has been at the forefront of this revolution. With the development of reusable rocket technology, such as the Falcon 9 and Falcon Heavy, SpaceX has dramatically reduced the cost of launching payloads into orbit. The company has also made significant progress towards its goal of colonizing Mars, with the development of the Starship, a fully reusable spacecraft designed for interplanetary travel.

The Emergence of Advanced Propulsion Technologies

As humanity looks towards the stars, the need for advanced propulsion technologies capable of efficient and rapid space travel has become increasingly apparent. Current chemical rocket propulsion systems are limited in their ability to propel spacecraft over vast distances in a timely and efficient manner.

Emerging propulsion technologies, such as ion thrusters, nuclear thermal propulsion, and solar sails, offer promising alternatives. These advanced systems have the potential to revolutionize space travel, enabling missions to distant planets, moons, and even other star systems.

The Future of Space Technology

The evolution of space technology has been a journey marked by progress, setbacks, and above all, human ingenuity. As we look to the future, the possibilities for space exploration and technology seem almost limitless. Some of the key areas of development likely to shape the future of space technology include:

1. Space tourism: Companies like Blue Origin and Virgin Galactic are working towards making space tourism a reality, offering suborbital flights for private citizens in the coming years.
2. Lunar bases and Mars colonization: Plans for establishing permanent human settlements on the Moon and Mars are underway, with organizations like NASA and SpaceX leading the charge.
3. Space mining: The extraction of valuable resources from celestial bodies, such as asteroids, could become a critical component of the space economy and a driver for further technological advancements.
4. Interstellar travel: The development of advanced propulsion systems and the prospect of discovering habitable exoplanets could one day enable humanity to embark on interstellar missions, further expanding our reach into the cosmos.

The evolution of space technology has been a remarkable testament to human curiosity, ambition, and innovation. From the early days of rocketry to the ongoing development of advanced propulsion systems and reusable spacecraft, we have come a long way in our understanding and exploration of the cosmos. As we continue to push the boundaries of what is possible, the future of space technology promises to open up new frontiers and unlock the potential for humanity to reach even greater heights.

86

The Future of Space Exploration

Space exploration has captured the imagination of people around the world for decades. From the first manned mission to the moon in 1969 to the recent landing of the Perseverance rover on Mars, space exploration has been at the forefront of scientific discovery and innovation. As we look to the future, it's clear that space exploration will continue to play a vital role in our understanding of the universe and our place within it. In this article, we will explore the future of space exploration and what it might look like in the years to come.

1. Continued Exploration of Mars

Mars has long been a focus of space exploration, and it's likely that this trend will continue in the years to come. In addition to the Perseverance rover, which is currently exploring the surface of Mars, NASA and other space agencies are planning a number of additional missions to the red planet. These missions could include the return of samples from Mars, the deployment of additional rovers and landers, and even the establishment of a human presence on the planet.

1. The Search for Life Beyond Earth

One of the most exciting aspects of space exploration is the search for life beyond Earth. Scientists believe that there may be other habitable planets in our galaxy, and a number of missions are currently underway to search for signs of life on these planets. For example, the James Webb Space Telescope, set to launch in 2021, will be able to analyze the atmospheres of exoplanets in greater detail than ever before, potentially allowing scientists to detect the presence of biosignatures and potentially confirm the existence of extraterrestrial life.

1. Advancements in Spacecraft Technology

As technology continues to improve, we can expect to see new and innovative spacecraft designs that are faster, more efficient, and more reliable than ever before. For example, NASA is currently developing the Space Launch System, which will be the most powerful rocket ever built, capable of taking humans to Mars and beyond. Private companies such as SpaceX and Blue Origin are also working on developing new spacecraft that could make space travel more accessible and affordable for the general public.

1. Increased Collaboration Among Space Agencies

Space exploration has historically been a competition between nations, but in recent years we've seen an increasing trend towards collaboration among space agencies. For example, NASA is working with the European Space Agency and the Japanese Aerospace Exploration Agency on the James Webb Space Telescope, and the International Space Station is a collaborative effort between the United States, Russia, Europe, Japan, and Canada. As the challenges of space exploration become more complex and expensive, it's likely that we will see more of this type of collaboration in the future.

1. The Commercialization of Space

In addition to government space agencies, private companies are also playing an increasingly important role in space exploration. Companies like SpaceX, Blue Origin, and Virgin Galactic are developing new spacecraft and technologies that could revolutionize the way we travel to space. In addition, these companies are also exploring the possibility of space tourism, which could make space exploration more accessible and affordable for the general public.

· **The Development of Space Mining**

As we continue to explore the universe, we're likely to discover new resources that could be valuable here on Earth. For example, asteroids and other celestial bodies are rich in minerals such as iron, nickel, and platinum. Companies are already exploring the possibility of space mining, which could provide a new source of valuable resources for our planet.

The future of space exploration is incredibly exciting, with new missions, technologies, and discoveries on the horizon. As we continue to explore the universe, we're likely to gain a better understanding of our place within it and make groundbreaking discoveries that could revolutionize our world.

International Cooperation and Space Law The exploration of space requires global cooperation.

The exploration of space is a global endeavor that requires international cooperation and coordination. Space missions involve multiple nations and require the sharing of resources, knowledge, and technology. As such, international cooperation is essential in achieving our goals of exploring the universe and understanding our place within it. In addition, the development of space law is also crucial in regulating the activities of nations and private enterprises in space. In this article, we will explore the role of international cooperation in space missions and the development of space law.

International Cooperation in Space Missions

International cooperation has been a key component of space exploration since the beginning of the space race. The International Space Station (ISS), for example, is a collaborative effort between the United States, Russia, Europe, Japan, and Canada. The ISS has been continuously occupied since 2000 and serves as a platform for scientific research and technological development.

In addition to the ISS, there have been many other examples of international cooperation in space missions. For example, the Hubble Space Telescope, which has provided us with stunning images of the universe, was a joint project between NASA and the European Space Agency. The Mars Exploration Rovers, Spirit and Opportunity, were also a joint project between NASA and the European Space Agency.

International cooperation in space missions is essential for a number of reasons. First, it allows for the sharing of resources and expertise, which can help to reduce costs and increase efficiency. Second, it can help to promote peace and understanding between nations, as countries work together towards a common goal. Finally, it can help to advance scientific knowledge and technological development, as nations share their research and development.

Development of Space Law

As space exploration and commercial activity in space increases, it is becoming increasingly important to develop space law to regulate the activities of nations and private enterprises in space. Space law is a branch of international law that deals with the legal framework governing space activities. It encompasses a wide range of issues, including space exploration, the use of space resources, and the regulation of commercial activities in space.

The Outer Space Treaty, which was signed in 1967, is the cornerstone of space law. It established the principles governing the exploration and use of outer space, including the peaceful use of outer space and the prohibition of weapons of mass destruction in space. The treaty has been ratified by over 100 countries, including all of the major spacefaring nations.

In addition to the Outer Space Treaty, there are a number of other international agreements and treaties that govern space activities. These include the Moon Agreement, which provides a framework for the use of the Moon and other celestial bodies, and the Liability Convention, which governs issues related to liability for damage caused by space activities.

As space exploration and commercial activity in space continue to grow, it is likely that we will see the development of new space laws and regulations. For example, there may be a need to develop regulations for space mining, which could involve the extraction of resources from asteroids and other celestial bodies.

International cooperation and the development of space law are essential in ensuring the safe and responsible exploration and use of outer space. As we continue to explore the universe and develop new technologies, it is important that we work together as a global community to advance our understanding of the cosmos and to regulate the activities of nations and private enterprises in space. By doing so, we can ensure that space remains a peaceful and cooperative frontier for generations to come.

87

The Secret of Time Travel

1. **The Nature of Time Before delving into time travel, it is crucial to understand the concept of time itself.**

Time is a fundamental aspect of our existence, and it plays a crucial role in our understanding of the world around us. However, despite its central importance, the nature of time remains a mystery in many ways. In this article, we will explore the concept of time and the theories that attempt to explain its nature.

The Arrow of Time

One of the most fundamental aspects of time is the arrow of time, which refers to the fact that time appears to have a direction or flow. We experience time as moving forward, with the past behind us and the future ahead. This is in contrast to other physical phenomena, such as the laws of physics, which appear to be symmetrical in time and do not have a preferred direction.

The arrow of time has been the subject of much debate and speculation among physicists and philosophers. Some theories suggest that the arrow of time is related to the expansion of the universe, with time moving forward as the universe expands. Others suggest that the arrow of time is related to the thermodynamic properties of matter, with time moving forward as the entropy of the universe increases.

Relativity

Another theory that has had a significant impact on our understanding of time is the theory of relativity. According to relativity, time is not an absolute quantity, but is instead relative to the observer. This means that the passage of time can be different for different observers, depending on their relative velocities or positions in space.

The theory of relativity has been tested and confirmed by numerous experiments, and has had a profound impact on our understanding of the universe. It has led to new insights into the nature of space and time, and has opened up new avenues for research in fields such as cosmology and particle physics.

Entropy and the Direction of Time

One of the most intriguing theories about the nature of time is the idea that it is defined by the direction of entropy. Entropy is a measure of the disorder or randomness of a system, and it has been suggested that the direction of time is related to the tendency of entropy to increase over time.

According to this theory, the past appears ordered and predictable because entropy was lower in the past, while the future appears uncertain and unpredictable because entropy is increasing. This theory has been supported by a wide range of observations and experiments, and has led to new insights into the nature of time and the universe as a whole.

The nature of time is a complex and fascinating subject, and it has been the subject of much debate and speculation among physicists, philosophers, and other thinkers throughout history. The theories we have discussed here, such as the arrow of time, relativity, and the role of entropy in defining the direction of time's flow, are just a few of the many ideas that have been put forward in attempts to explain this fundamental aspect of our existence. As we continue to explore the universe and develop new technologies, it is likely that our understanding of time will continue to evolve, leading to new insights and discoveries that could change the way we think about ourselves and the world around us.

2.Time Travel in Physics The possibility of time travel has long fascinated scientists and the general public alike.

The concept of time travel has captivated the imaginations of people for centuries, with countless stories, films, and TV shows exploring the idea of traveling through time. However, while time travel may seem like a fantastical idea, it is a topic that has been explored by physicists and scientists for decades. In this article, we will examine the various theories of time travel in physics, including the role of wormholes, black holes, and the concept of closed timelike curves.

Wormholes

One of the most popular ideas for time travel involves the use of wormholes, which are theoretical objects that could connect two separate points in space-time. The idea is that if we could create or find a wormhole, we could use it to travel from one point in time to another, essentially bypassing the time in between.

However, while wormholes are theoretically possible according to the laws of physics, there are many challenges to actually using them for time travel. For example, it is not clear how one would actually create or stabilize a wormhole, and there are questions about whether it would be safe to use for time travel, as entering a wormhole could involve passing through intense gravitational forces and radiation.

Black Holes

Another idea for time travel involves the use of black holes, which are extremely dense objects with incredibly strong gravitational forces. According to the theory of relativity, if one were to travel close enough to a black hole, time would slow down relative to the rest of the universe. This means that if someone were to spend a significant amount of time near a black hole and then return to the rest of the universe, they would have effectively traveled forward in time.

However, while black holes may offer a theoretical possibility for time travel, there are many challenges to actually using them for this purpose. For example, black holes are incredibly dangerous and inhospitable environments, and it is not clear how one would safely travel near one.

Closed Timelike Curves

Finally, the concept of closed timelike curves is another idea for time travel. According to this theory, if an object were to travel in a closed loop through space-time, it could theoretically travel back in time and encounter itself in the past. This idea has been explored in a number of thought experiments, but it is not clear whether closed timelike curves are physically possible.

While the idea of time travel has been explored in countless works of fiction, it remains a topic of great interest and fascination in the scientific community as well. While the theories we have discussed here offer some possible mechanisms for time travel, they are all highly speculative and face significant challenges and obstacles. It remains to be seen whether time travel will ever be possible in the real world, but the exploration of these ideas and theories is an important part of scientific inquiry and imagination.

3.Paradoxes and Consequences Time travel presents numerous paradoxes and challenges.

Time travel is a fascinating concept that has captured the imaginations of people for centuries. However, the idea of traveling through time also presents numerous paradoxes and challenges that have been the subject of much debate and speculation. In this article, we will delve into some of the most famous thought experiments and paradoxes related to time travel, including the grandfather paradox, the bootstrap paradox, and the implications of changing the past.

The Grandfather Paradox

Perhaps the most famous paradox associated with time travel is the grandfather paradox. This paradox is based on the idea that if someone were to travel back in time and kill their own grandfather before he had children, they would never be born, which would mean they could never travel back in time to kill their grandfather in the first place.

The grandfather paradox highlights the potential consequences of changing the past and the complexities of time travel. It also raises important questions about the nature of causality and the interconnectedness of events in time.

The Bootstrap Paradox

Another famous paradox associated with time travel is the bootstrap paradox, which involves the idea of an object or piece of information being its own origin. For example, imagine someone travels back in time and gives Beethoven a copy of his own sheet music. Beethoven then uses the sheet music to compose his famous symphonies, which the time traveler later takes back to the future and publishes under his own name.

The bootstrap paradox highlights the potential for time travel to create causal loops and the difficulties in assigning origins to objects or information that appear to have no beginning.

Changing the Past

The idea of changing the past is one of the most intriguing aspects of time travel. However, it also presents a number of challenges and consequences. For example, if someone were to travel back in time and change a key event, such as preventing a war or saving a life, they could potentially alter the course of history and create a butterfly effect of unintended consequences.

The implications of changing the past also raise important questions about free will and determinism, and the role of individual actions in shaping the course of history.

Time travel presents numerous paradoxes and challenges, from the potential consequences of changing the past to the complexities of causality and origins. While the concept of time travel remains largely speculative, the exploration of these paradoxes and thought experiments offers important insights into the nature of time, causality, and the interconnectedness of events. As our understanding of the universe continues to evolve, it is likely that we will continue to explore these ideas and theories, pushing the boundaries of scientific inquiry and imagination.

4.Practical Limitations and Ethical Considerations While the idea of time travel is tantalizing, there are several practical and ethical concerns to

consider.

The concept of time travel has been the subject of much fascination and speculation throughout history. However, while the idea of traveling through time may seem appealing, there are several practical and ethical concerns that must be considered. In this article, we will discuss the technological hurdles that must be overcome and the ethical dilemmas surrounding the potential manipulation of time.

Technological Hurdles

One of the most significant practical limitations of time travel is the technological challenges involved in building a time machine. While there are many theories about how time travel might be possible, such as through the use of wormholes or black holes, the technology required to create such devices is currently beyond our capabilities.

Even if we were to develop the necessary technology, there would likely be significant practical challenges involved in actually using a time machine. For example, time travel could involve passing through intense gravitational fields or radiation, which could pose significant health risks.

Ethical Considerations

In addition to the technological hurdles involved in time travel, there are also a number of ethical dilemmas to consider. One of the most significant concerns is the potential for manipulation of time and the impact that could have on the course of history.

For example, if someone were to travel back in time and change a key event, such as preventing a war or altering the outcome of an election, they could potentially create a butterfly effect of unintended consequences that could dramatically alter the course of history.

There are also concerns about the potential for time travel to be used for personal gain or to interfere with the lives of others. For example, someone could use time travel to accumulate wealth or power, or to manipulate the actions of others for their own benefit.

While the idea of time travel may seem appealing, there are several practical and ethical considerations that must be taken into account. The technological hurdles involved in building a time machine are significant, and there are also concerns about the potential for manipulation of time and the impact that could have on the course of history. As we continue to explore the mysteries of the universe and push the boundaries of scientific inquiry, it is important that we consider these practical and ethical considerations to ensure that any advances we make are used for the greater good.

5.Time Travel in Popular Culture Time travel has been a popular theme in literature, film, and television

Time travel has long captured the imaginations of people, and it has been a popular theme in literature, film, and television for many years. From classic stories like H.G. Wells' "The Time Machine" to contemporary films like "Avengers: Endgame," time travel has been used to explore a wide range of themes and ideas. In this article, we will explore some of the most famous and influential time travel stories in popular culture, examining how they have shaped our understanding of this enigmatic concept.

"The Time Machine" by H.G. Wells

One of the earliest and most influential works of time travel fiction is H.G. Wells' "The Time Machine," published in 1895. In this classic novel, a time traveler travels to the year 802,701 AD and discovers a world divided into two groups: the Eloi and the Morlocks. The story explores themes of evolution, social class, and the potential consequences of humanity's actions.

"The Time Machine" has had a significant impact on our understanding of time travel and its possibilities, and it has influenced countless works of science fiction and fantasy that have followed.

"Back to the Future" Trilogy

Perhaps the most iconic time travel story in modern popular culture is the "Back to the Future" trilogy, which follows the adventures of Marty McFly and Doc Brown as they travel through time in a modified DeLorean. The films explore themes of destiny, the butterfly effect, and the importance of personal responsibility.

The "Back to the Future" trilogy has become a cultural touchstone, and it has had a significant impact on our understanding of time travel and its possibilities. The films have also inspired numerous parodies, homages, and tributes in popular culture.

"Doctor Who"

Another iconic time travel story in popular culture is "Doctor Who," a long-running British television series that first aired in 1963. The show follows the adventures of the Doctor, a time-traveling alien who travels through time and space in his TARDIS, a spaceship disguised as a blue police box.

"Doctor Who" has explored a wide range of themes and ideas over the years, from the nature of time and the consequences of meddling with history to the importance of compassion and empathy. The show has had a significant impact on popular culture, and it has inspired numerous spin-offs, merchandise, and adaptations.

"Avengers: Endgame"

One of the most recent and influential time travel stories in popular culture is "Avengers: Endgame," the 2019 superhero film that concluded the Marvel Cinematic Universe's Infinity Saga. The film follows the Avengers as they attempt to undo the damage caused by the villainous Thanos by traveling back in time to retrieve the Infinity Stones.

"Avengers: Endgame" explores themes of loss, sacrifice, and the importance of teamwork, and it has had a significant impact on popular culture. The film has also sparked numerous debates and discussions among fans and critics about the nature of time travel and its implications.

Time travel has been a popular theme in literature, film, and television for many years, and it has had a significant impact on popular culture. From classic stories like "The Time Machine" to contemporary films like "Avengers: Endgame," time travel has been used to explore a wide range of themes and ideas, from the consequences of humanity's actions to the importance of personal responsibility. As we continue to explore the mysteries of the universe and push the boundaries of scientific inquiry, it is likely that time travel will continue to captivate our imaginations and inspire new stories and ideas.

88

The Mystery of Parallel Universes

The idea of parallel universes has captivated the human imagination for centuries. It suggests the possibility that there are other versions of ourselves living in alternate realities, and that every decision we make leads to different outcomes in different universes. While the concept may seem like science fiction, it is actually based on real scientific theories, such as the many-worlds interpretation of quantum mechanics and the cosmic inflation theory of the early universe.

The Many-Worlds Interpretation

The many-worlds interpretation of quantum mechanics suggests that every time a quantum event occurs, the universe splits into multiple parallel universes, each corresponding to a different outcome of the event. For example, if you flip a coin, there would be one universe in which it landed heads up and another in which it landed tails up. While this interpretation is just one of many theories of the multiverse, it has gained significant attention due to its implications for the nature of reality and the possibility of time travel.

The Cosmic Inflation Theory

Another theory of the multiverse is the cosmic inflation theory, which proposes that the universe underwent a period of rapid expansion shortly after the Big Bang. During this expansion, small quantum fluctuations were amplified to create large-scale structures in the universe, including galaxies and clusters of galaxies. According to this theory, there may be other regions of space beyond what we can observe, each with its own set of physical laws and properties.

Implications and Controversies

The concept of parallel universes raises many questions and controversies in the scientific community. Some scientists argue that there is no direct evidence to support the existence of parallel universes, and that the theories are purely speculative. Others argue that the idea is grounded in scientific principles and that it may help us better understand the nature of the universe and the fundamental laws that govern it.

One potential implication of parallel universes is the possibility of time travel. If every decision we make leads to different outcomes in different universes, it could be possible to travel to a parallel universe where a different decision was made. However, this theory raises questions about the possibility of changing the past and the potential for paradoxes.

The Multiverse Hypothesis

The multiverse hypothesis is a theoretical framework that posits the existence of multiple universes, each with its own set of physical laws and properties. This idea has its roots in several areas of theoretical physics, including quantum mechanics, string theory, and cosmic inflation.

1. Quantum Mechanics: In quantum mechanics, particles exist in a superposition of states until they are measured, at which point they collapse into a single state. The Many Worlds Interpretation (MWI) of quantum mechanics, first proposed by Hugh Everett III in 1957, suggests that every time a quantum measurement is made, the universe splits into separate branches, each representing a different outcome of the measurement. According to this interpretation, there are countless parallel universes in which every possible quantum outcome is realized.

2. String Theory: String theory, a leading candidate for a unified theory of everything, proposes that all particles and forces in the universe arise from tiny, vibrating strings. In certain formulations of string

theory, the existence of multiple universes is a natural consequence of the underlying mathematics. These "brane worlds" are said to exist in higher-dimensional spaces and may interact with one another through gravity or other forces.

3. Cosmic Inflation: Cosmic inflation is a widely accepted theory that explains the uniformity and flatness of the observable universe. According to this theory, the universe underwent a brief period of rapid expansion in the moments following the Big Bang. Some physicists, such as Alan Guth and Andrei Linde, have proposed that inflation may have given rise to multiple, causally disconnected "bubble universes," each with its own physical properties and initial conditions.

The Many-Worlds Interpretation

One of the most well-known theories of the multiverse is the many-worlds interpretation of quantum mechanics. This theory proposes that every time a quantum event occurs, the universe splits into multiple parallel universes, each corresponding to a different outcome of the event.

For example, imagine a coin toss. In our universe, the coin will either land heads up or tails up. However, according to the many-worlds interpretation, each possible outcome corresponds to a separate universe. In one universe, the coin lands heads up, while in another universe, it lands tails up.

While the many-worlds interpretation is just one of many theories of the multiverse, it has gained significant attention in recent years due to its implications for the nature of reality and the possibility of time travel.

The Cosmic Landscape

Another theory of the multiverse is the cosmic landscape, which suggests that there are many different possible configurations of the universe, each with its own set of physical laws and properties. This theory is based on the concept of string theory, which proposes that the fundamental particles of the universe are actually tiny strings of energy.

In the cosmic landscape, the different configurations of the universe correspond to different possible values of the parameters of string theory. Each possible configuration is thought to correspond to a different parallel universe, each with its own unique properties.

Quantum Mechanics and Parallel Universes

Quantum mechanics plays a significant role in the theory of parallel universes. One of the key concepts in quantum mechanics is quantum superposition, which suggests that particles can exist in multiple states simultaneously until they are observed or measured.

This concept is closely related to the idea of parallel universes, as it suggests that multiple possible outcomes of a quantum event can exist simultaneously. The implications of quantum superposition are perhaps best illustrated by Schrödinger's cat thought experiment, in which a cat in a sealed box can be considered to be both alive and dead until the box is opened and the cat is observed.

The Implications of Parallel Universes

The concept of parallel universes, or the existence of multiple alternate realities, has implications that extend far beyond the realm of science fiction. While the idea may seem speculative, it is grounded in scientific principles and has the potential to shed light on some of the most profound mysteries of the universe. In this article, we will explore some of the implications of parallel universes, including the possibility of time travel, the nature of reality, and the limits of scientific understanding.

Time Travel

One of the most intriguing implications of parallel universes is the possibility of time travel. According to the many-worlds interpretation of quantum mechanics, every time a quantum event occurs, the universe splits into multiple parallel universes, each corresponding to a different outcome of the event. This suggests that every decision we make creates a new parallel universe in which a different decision was made.

If time travel were possible, it could be used to travel between parallel universes and explore the alternate realities that exist within them. However, this theory raises many questions and paradoxes, such as the possibility of changing the past and creating a new future.

The Nature of Reality

The concept of parallel universes also has implications for our understanding of the nature of reality. If there are multiple parallel universes, each with its own set of physical laws and properties, it raises questions about what is truly real and what is simply a product of our own perception.

Furthermore, the idea of parallel universes suggests that reality may be more complex and nuanced than we currently understand. It raises questions about the limits of scientific understanding and the possibility that there may be aspects of the universe that are beyond our ability to comprehend.

The Limits of Scientific Understanding

The concept of parallel universes also has implications for the limits of scientific understanding. While science has made tremendous progress in understanding the fundamental laws that govern the universe, there may be aspects of the universe that are beyond our ability to observe or measure.

The existence of parallel universes suggests that there may be other regions of space beyond what we can observe, each with its own set of physical laws and properties. It raises questions about the limits of scientific inquiry and the possibility that there may be fundamental aspects of the universe that are forever beyond our understanding.

The idea of parallel universes is a complex and fascinating topic that has captivated scientists and the general public alike. While the concept may seem like science fiction, it is grounded in real scientific theories, such as the many-worlds interpretation and the cosmic inflation theory. While the existence of parallel universes remains a topic of debate and speculation, it continues to inspire new ideas and discoveries in the field of theoretical physics.

89

The Power of Nanotechnology

1. **Introduction to Nanotechnology Nanotechnology involves manipulating matter at the atomic and molecular scale.**

Nanotechnology is a rapidly evolving field that involves manipulating matter at the atomic and molecular scale. This exciting area of research has the potential to revolutionize many aspects of modern life, from medicine to energy production. In this article, we will provide an overview of nanotechnology, its history, and its potential applications in various fields.

What is Nanotechnology?

Nanotechnology involves the manipulation of materials at the nanoscale, which is typically defined as being between 1 and 100 nanometers in size. At this scale, the properties of materials can be significantly different from their bulk counterparts, allowing for the development of new materials and devices with unique properties and capabilities.

Nanotechnology is a highly interdisciplinary field, drawing on concepts and techniques from physics, chemistry, biology, and engineering. It has the potential to impact a wide range of fields, including medicine, electronics, energy production, and environmental science.

History of Nanotechnology

The concept of manipulating matter at the nanoscale was first proposed by physicist Richard Feynman in his 1959 talk "There's Plenty of Room at the Bottom." However, it wasn't until the development of the scanning tunneling microscope in the 1980s that researchers were able to directly observe and manipulate individual atoms and molecules.

Since then, advances in nanotechnology have been rapid, with new materials and devices being developed at an ever-increasing pace. In 2000, the U.S. government launched the National Nanotechnology Initiative to promote research and development in the field, and today, nanotechnology is a major area of research worldwide.

Applications of Nanotechnology

Nanotechnology has the potential to impact a wide range of fields, including:

Medicine: Nanotechnology is being used to develop new treatments for cancer and other diseases, as well as new diagnostic tools and drug delivery systems.

Electronics: Nanotechnology is being used to develop new materials and devices for electronics, such as nanoscale transistors and sensors.

Energy production: Nanotechnology is being used to develop new materials and devices for energy production and storage, such as nanoscale solar cells and batteries.

Environmental science: Nanotechnology is being used to develop new materials and devices for pollution control and remediation, such as nanoscale filters and sensors.

Nanotechnology is a rapidly evolving field that has the potential to revolutionize many aspects of modern life. Its interdisciplinary nature and potential applications in fields such as medicine, electronics, energy production, and environmental science make it an exciting area of research with significant potential for impact. As research in nanotechnology continues to progress, we can expect to see new materials, devices, and applications emerge, shaping the way we live and work in the years to come.

2. Nanomaterials and Nanodevices At the heart of nanotechnology are nanomaterials and nanodevices.

Nanotechnology involves manipulating matter at the nanoscale, which is typically defined as being between 1 and 100 nanometers in size. At this scale, the properties of materials can be significantly different from their bulk counterparts, allowing for the development of new materials and devices with unique properties and capabilities. In this article, we will explore the different types of nanomaterials and nanodevices that are at the heart of nanotechnology.

Types of Nanomaterials

There are many different types of nanomaterials, each with unique properties and potential applications. Some of the most commonly studied nanomaterials include:

Carbon nanotubes: These are long, cylindrical tubes made of carbon atoms that are arranged in a unique way. Carbon nanotubes have exceptional strength and conductivity, and they are being studied for use in electronics, energy storage, and other applications.

Graphene: Graphene is a single layer of carbon atoms arranged in a hexagonal lattice. It is an extremely strong and conductive material that is being studied for use in electronics, energy storage, and other applications.

Quantum dots: These are tiny semiconductor particles that exhibit unique optical and electronic properties. They are being studied for use in solar cells, displays, and other applications.

Nanodevices

Nanotechnology is also being used to develop new types of nanodevices, which are tiny machines or sensors that operate at the nanoscale. Some of the most promising nanodevices being developed include:

Nanosensors: These are sensors that can detect and measure things like temperature, pressure, and chemical concentrations at the nanoscale. They are being studied for use in medical diagnostics, environmental monitoring, and other applications.

Nanorobots: These are tiny machines that can be programmed to perform specific tasks, such as delivering drugs to specific cells in the body. They are being studied for use in medicine and other applications.

Nanoelectronics: These are electronic devices that operate at the nanoscale, such as transistors and memory devices. They are being studied for use in computing and other applications.

Conclusion

Nanomaterials and nanodevices are at the heart of nanotechnology, offering unique properties and capabilities that can be harnessed for a wide range of applications. From carbon nanotubes and graphene to quantum dots and nanosensors, there are many different types of nanomaterials being studied for their potential applications in electronics, energy production, medicine, and environmental monitoring. As research in nanotechnology continues to progress, we can expect to see new materials and devices emerge, shaping the way we live and work in the years to come.

3.Medical Applications of Nanotechnology One of the most promising areas for nanotechnology is its application in medicine.

Nanotechnology has the potential to revolutionize many fields, including medicine. At the heart of this promise is the ability to manipulate matter at the nanoscale, allowing for the creation of new materials and devices with unique properties and capabilities. In this article, we will explore the use of nanotechnology in medicine, focusing on its potential applications in drug delivery, tissue engineering, and diagnostics.

Drug Delivery

One of the most promising applications of nanotechnology in medicine is drug delivery. Nanoparticles can be designed to target specific cells or tissues in the body, allowing for more precise delivery of drugs and reducing the risk of side effects.

For example, nanoparticles can be coated with specific molecules that allow them to bind to cancer cells. This allows for targeted drug delivery to cancer cells, while minimizing the impact on healthy cells. Nanoparticles can also be used to deliver drugs across the blood-brain barrier, which can be a significant obstacle in the treatment of neurological diseases.

Tissue Engineering

Another area where nanotechnology shows great promise is in tissue engineering. Nanomaterials can be used to create scaffolds that mimic the structure of natural tissues, allowing for the growth of new tissues and organs.

For example, researchers have developed nanomaterials that can be used to regenerate bone tissue. These materials can be implanted into the body and stimulate the growth of new bone tissue, which can be used to treat injuries and diseases like osteoporosis.

Diagnostics

Nanotechnology is also being used to develop new diagnostic tools that can detect diseases at an early stage and monitor the effectiveness of treatments. Nanoparticles can be used to create highly sensitive sensors that can detect small changes in biological markers, allowing for the early detection of diseases like cancer.

For example, researchers have developed a nanoscale sensor that can detect proteins associated with pancreatic cancer. This could potentially allow for earlier diagnosis and more effective treatment of this deadly disease.

Conclusion

Nanotechnology has the potential to revolutionize many aspects of medicine, from drug delivery and tissue engineering to diagnostics and monitoring. The ability to manipulate matter at the nanoscale offers unique properties and capabilities that can be harnessed for a wide range of medical applications. As research in nanotechnology continues to progress, we can expect to see new materials, devices, and therapies emerge, shaping the way we treat and manage diseases in the years to come.

4.Environmental and Energy Applications Nanotechnology also holds the key to solving some of the world's most pressing environmental and energy challenges.

Nanotechnology holds immense potential in addressing some of the world's most pressing environmental and energy challenges. By manipulating matter at the nanoscale, researchers are developing new materials and devices that can help create a cleaner, more sustainable future. In this article, we will explore the use of nanotechnology in environmental and energy applications, focusing on water purification, air quality improvement, and renewable energy production.

Water Purification

Access to clean water is a critical global challenge. Nanotechnology offers promising solutions to this problem by providing a way to remove contaminants from water more efficiently and effectively than traditional methods.

Nanoparticles can be used as filters to remove impurities from water. They can also be used to create membranes that selectively allow water molecules to pass through while blocking contaminants. This can be particularly useful in desalination, where seawater can be filtered to produce fresh water.

Nanotechnology can also be used to create materials that remove specific contaminants from water. For example, nanoparticles can be coated with specific molecules that allow them to bind to heavy metals, toxins, or bacteria in water, removing them from the water supply.

Air Quality Improvement

Air pollution is another significant global challenge that can be addressed through the use of nanotechnology. Nanoparticles can be used to remove pollutants from the air, making it safer and healthier to breathe.

One application of nanotechnology in air quality improvement is the use of nanofilters. Nanofilters are designed to capture pollutants such as fine particulate matter, which can cause respiratory problems and other health issues. Nanoparticles can also be used to remove harmful gases, such as nitrogen oxides and sulfur dioxide, from the air.

Renewable Energy Production

Nanotechnology is also being used to develop new materials and devices for renewable energy production. By creating materials with unique properties and capabilities, researchers can improve the efficiency and effectiveness of renewable energy technologies.

For example, nanotechnology is being used to create more efficient solar cells. By using nanoparticles to increase the surface area of solar cells, researchers can capture more sunlight and generate more energy. Nanoparticles can also be used to create new materials for energy storage, such as batteries and supercapacitors.

Nanotechnology offers promising solutions to some of the world's most pressing environmental and energy challenges. By manipulating matter at the nanoscale, researchers are developing new materials and devices that can help create a cleaner, more sustainable future. From water purification and air quality improvement to renewable energy production, nanotechnology has the potential to make a significant impact on our planet and the way we live. As research in nanotechnology continues to progress, we can expect to see new materials, devices, and technologies emerge, shaping the way we address environmental and energy challenges in the years to come.

5. **Ethical and Societal Implications As with any emerging technology, nanotechnology raises ethical and societal concerns.**

Nanotechnology is a rapidly evolving field that has the potential to revolutionize many aspects of modern life. However, as with any emerging technology, it also raises ethical and societal concerns. In this article, we will explore the potential risks and ethical considerations surrounding nanotechnology, as well as the need for regulatory frameworks to ensure responsible development and use of this technology.

Environmental Impact

One of the most significant concerns surrounding nanotechnology is its potential environmental impact. Nanoparticles can be difficult to control and may have unintended consequences on ecosystems and human health. For example, nanoparticles could accumulate in the environment and disrupt natural processes or cause harm to organisms.

Additionally, the production and disposal of nanomaterials could contribute to pollution and waste, raising concerns about the sustainability of this technology. To address these concerns, it is essential to develop environmentally sustainable practices and ensure that the potential environmental impact of nanotechnology is carefully considered.

Privacy Concerns

Nanotechnology has the potential to create new surveillance and monitoring capabilities, raising concerns about privacy. For example, nanosensors could be used to monitor individuals' health or location, potentially violating their privacy rights.

It is important to consider how nanotechnology could be used in ways that infringe on individual rights and freedoms. Regulatory frameworks that balance innovation with ethical considerations must be established to ensure that the development and use of nanotechnology align with societal values.

Weaponization

Like any technology, nanotechnology has the potential to be weaponized. Nanoparticles could be developed for use in chemical or biological warfare, or for covert surveillance and monitoring.

Regulatory frameworks must be established to ensure that the development and use of nanotechnology is transparent, and that its potential for weaponization is mitigated. International collaboration and regulation are critical to ensure that nanotechnology is not used in ways that threaten global security and stability.

Nanotechnology is a powerful tool that has the potential to transform many aspects of modern life. However, as with any emerging technology, it also raises ethical and societal concerns. Environmental impact, privacy concerns, and the potential for weaponization are just a few of the issues that must be carefully considered as nanotechnology continues to evolve.

Regulatory frameworks that balance innovation with ethical consider-ations must be established to ensure that the development and use of nanotechnology align with societal values. International collaboration and regulation are critical to ensure that nanotechnology is not used in ways that threaten global security and stability.

By addressing these ethical and societal implications, we can ensure that nanotechnology is developed and used responsibly, contributing to a brighter, more sustainable future for all.

90

The Enigma of Transhumanism

Transhumanism is an enigmatic and controversial movement that aims to enhance the human condition through advanced technologies. Proponents argue that it offers boundless possibilities to improve our lives, while critics contend that it may lead to unforeseen consequences and ethical dilemmas. This article examines the multifaceted nature of transhumanism, exploring its potential benefits, drawbacks, and ethical implications.

· **The Promise of Transhumanism**

Transhumanism is rooted in the belief that we can use technology to overcome our biological limitations. It envisions a future where humans merge with machines, allowing us to live longer, healthier lives, and enhancing our cognitive and physical capabilities. The movement has several potential advantages:

A. Physical Enhancement: Prosthetics and exoskeletons are being developed to restore or augment human functionality. Bionics and implants can enhance sensory perception, and gene editing techniques like CRISPR can potentially eliminate genetic disorders.

B. Cognitive Enhancement: Researchers are exploring the development of brain-computer interfaces (BCIs) that could enable direct communication between our minds and digital devices. These technologies may help us to process information faster and more efficiently.

C. Life Extension: Breakthroughs in regenerative medicine, such as stem cell therapies and tissue engineering, may enable us to extend human lifespan and improve our overall health.

· The Dangers of Transhumanism

Despite its potential benefits, transhumanism also raises several concerns:

A. Ethical Dilemmas: Genetic engineering and cognitive enhancements may result in unintended consequences or exacerbate social inequalities. For example, access to these technologies could be limited to the wealthy, further widening the gap between the haves and have-nots.

B. Loss of Humanity: Critics argue that altering our biological makeup could lead to a loss of what makes us human. We may risk eroding the unique qualities that define our species, such as empathy, creativity, and emotional intelligence.

C. Unforeseen Consequences: As with any groundbreaking technology, there is the potential for unforeseen consequences. The development of advanced AI, for instance, could lead to a loss of control over these powerful systems or even result in an existential threat to humanity.

· Navigating the Ethical Landscape

As we venture into the uncharted territory of transhumanism, it is crucial that we approach these advancements with caution and ethical consideration. Some key principles to guide our exploration include:

A. Inclusivity: Ensuring that access to these technologies is equitable and available to all, regardless of socioeconomic status.

B. Transparency: Promoting open and honest dialogue about the potential risks and benefits of transhumanist technologies, fostering informed decision-making.

C. Regulation: Establishing responsible governance and oversight mechanisms to safeguard against unintended consequences and ensure the ethical use of these technologies.

Transhumanism represents both an opportunity and a challenge. On one hand, it offers the potential to transcend our biological limitations and usher in a new era of human evolution. On the other hand, it raises complex ethical questions and potential dangers that must be carefully considered. As we grapple with the enigma of transhumanism, it is essential that we proceed with caution, open-mindedness, and a commitment to ethical decision-making.

X

Secrets of Conspiracy Theories

91

The Secret of the New World Order

Conspiracy theories have always been a part of human history, with people speculating about hidden agendas, secret societies, and plots to control the world. One of the most enduring and controversial conspiracy theories is the New World Order.

The New World Order conspiracy theory claims that a powerful global elite is planning to create a new world order, which would be a totalitarian government that would rule the world with an iron fist. According to the theory, this elite is made up of powerful politicians, businessmen, and bankers who are secretly working together to achieve their goals.

Proponents of the New World Order theory point to various pieces of evidence to support their claims. They cite the influence of organizations like the Bilderberg Group, the Trilateral Commission, and the Council on Foreign Relations, which they believe are working to create a global government. They also point to the symbols and imagery used on the US dollar bill, which they claim reveal the hidden agendas of the elite.

However, most of these claims have been debunked by experts in politics and history. The Bilderberg Group, for example, is a private organization that meets annually to discuss global issues, but it has no power to create or enforce policies. Similarly, the symbols on the US dollar bill are simply part of the design and have no hidden meaning.

Despite the lack of evidence to support the New World Order theory, it continues to have a strong following among some groups of people. Some believe that the theory is part of a larger conspiracy to distract people from the real issues facing society, while others see it as a warning about the dangers of unchecked government power.

The prevalence of the New World Order theory is a reminder of the power of conspiracy theories to capture people's imaginations and shape their beliefs. While some theories may have a grain of truth to them, it is important to approach them with a critical eye and rely on reputable sources of information to understand complex issues. Ultimately, the secrets of conspiracy theories may reveal more about human psychology and our need for meaning and purpose than about the hidden agendas of powerful elites.

92

The Mystery of Chemtrails

Chemtrails, also known as "contrails," are the visible trails left behind by planes flying at high altitudes. While many people believe that these trails are the result of chemicals being sprayed into the atmosphere for nefarious purposes, the scientific consensus is that they are simply a byproduct of jet engines.

Contrails are formed when hot exhaust gases from the plane's engines mix with the cold air at high altitudes, causing water vapor to condense into ice crystals. These ice crystals then form the visible white streaks that we see in the sky.

However, some conspiracy theorists claim that the trails left behind by planes are actually part of a secret government program to control the weather, manipulate the population, or even spread harmful chemicals into the air. They believe that these "chemtrails" are part of a vast and sinister plot to control the world.

Despite these claims, there is no credible evidence to support the idea that chemtrails are anything other than contrails. Scientists and government agencies have conducted numerous studies and found no evidence of any chemical spraying program. In fact, the idea of chemtrails has been thoroughly debunked by the scientific community.

While the origins of the chemtrails conspiracy theory are unclear, it has gained a following among some groups of people who distrust the government and believe in various conspiracy theories. The spread of misinformation about chemtrails has led to fear and anxiety among some people, who may believe that they are being exposed to harmful chemicals or that the government is hiding something from them.

In reality, the mystery of chemtrails is not a mystery at all. They are simply a natural phenomenon caused by the interaction of jet engines with the atmosphere. It is important to rely on credible sources of information and scientific evidence when evaluating claims about conspiracy theories or other controversial topics. While it is natural to be curious about the world around us, it is also important to approach these questions with a critical eye and a healthy dose of skepticism.

93

The Enigma of the Mandela Effect

The Mandela Effect is a phenomenon that has captured the imagination of many people in recent years. It refers to the collective misremembering of certain events, names, or details, with many people believing that their version of reality is the correct one. The name "Mandela Effect" comes from the belief that many people remember Nelson Mandela dying in prison in the 1980s, even though he was actually released and lived until 2013.

One of the most famous examples of the Mandela Effect is the children's book series, The Berenstain Bears. Many people remember the name of the family being spelled as "Berenstein," with an "e" instead of an "a." However, all evidence shows that the correct spelling is "Berenstain."

Other examples of the Mandela Effect include people misremembering movie quotes, logos, and even the placement of geographic locations on a map. Some people believe that these misrememberings are evidence of alternate universes or parallel realities, while others attribute them to faulty memory or suggestion.

Psychologists who have studied the Mandela Effect believe that it is likely caused by a combination of factors, including the fallibility of human memory, the power of suggestion, and the spread of misinformation through social media and other channels. It is also possible that the Mandela Effect is a result of the brain's tendency to fill in gaps in information or to create patterns where none exist.

The enigma of the Mandela Effect raises questions about the nature of reality and the fallibility of human perception. It is a reminder that our memories can be unreliable and that our understanding of the world around us is not always as clear-cut as we might like to think. While the Mandela Effect may seem like a strange and mysterious phenomenon, it is ultimately a testament to the complexity and richness of human experience.

94

The Power of the Deep State

The concept of the Deep State has gained increasing attention in recent years, especially in the United States. It refers to the idea that there is a shadowy, unelected group of people who hold significant power and influence over the government and society as a whole. While the existence of a Deep State is often debated, the concept raises important questions about power, democracy, and the role of institutions in society.

Defining the Deep State

The term "Deep State" is often used to describe a wide range of institutions and organizations, including the military, intelligence agencies, and other parts of the government bureaucracy. It is often characterized as a hidden, parallel government that operates behind the scenes and wields significant power and influence.

The concept of the Deep State has its roots in the Turkish political system, where it has been used to describe a network of military and intelligence officials who have held significant power and influence over the government for decades. In the United States, the concept gained increased attention following the 2016 presidential election, with some people suggesting that a Deep State was working to undermine the administration of President Trump.

Critics of the concept of the Deep State argue that it is a conspiracy theory that lacks any real evidence. They point out that democratic institutions and checks and balances exist to prevent any one group from holding too much power. However, proponents of the Deep State argue that these institutions are often co-opted or corrupted by powerful interests, leading to the emergence of a hidden power structure.

The Power of the Deep State

One of the key features of the Deep State is its ability to exert power and influence over government and society without being held accountable to democratic institutions or the public. This can take many forms, including the ability to manipulate the media, control access to information, and shape public opinion.

In some cases, the Deep State may work to undermine elected officials or policies that are seen as a threat to its interests. This can take the form of leaks to the media, bureaucratic stonewalling, or even outright sabotage.

Another key aspect of the Deep State is its ability to operate outside of the law. This can include illegal surveillance, extrajudicial killings, and other activities that are not subject to democratic oversight or accountability.

The existence of a Deep State raises important questions about the balance of power in society and the role of democratic institutions in ensuring that power is held accountable. It also highlights the potential dangers of allowing unelected officials and institutions to hold significant power and influence over government and society.

Challenges to Democracy

One of the main criticisms of the concept of the Deep State is that it undermines the democratic process by suggesting that there is a hidden power structure that is not subject to democratic accountability. This can lead to a sense of disillusionment and apathy among the public, who may feel that their voices do not matter and that the system is rigged against them.

However, proponents of the Deep State argue that the concept is necessary to hold powerful institutions and individuals accountable for their actions. They argue that the existence of a Deep State is a sign that democratic institutions are failing to hold power to account and that it is necessary to expose and dismantle these structures in order to restore democracy.

Ultimately, the challenge of the Deep State is to strike a balance between the need for accountability and transparency and the need for stability and continuity in government and society. It is a reminder that power can be wielded in many different ways and that the institutions of democracy must be constantly vigilant in ensuring that power is held accountable and that the voices of the people are heard.

The concept of the Deep State is a complex and controversial one, with many different interpretations and perspectives. While some see it as a dangerous conspiracy theory, others view it as a necessary tool for holding powerful interests to account.

95

The Mystery of the Denver Airport

The Denver International Airport (DIA) is one of the largest and busiest airports in the world, handling millions of passengers every year. However, the airport is also the subject of a number of conspiracy theories and urban legends that suggest that there is more going on than meets the eye.

One of the most persistent rumors about DIA is that it was built on top of an underground bunker or city that is used by the government or other organizations for secret purposes. Some people believe that this bunker is intended to serve as a refuge in the event of a catastrophic event, while others believe that it is used for more nefarious purposes, such as conducting experiments on unsuspecting individuals or hiding evidence of extraterrestrial life.

Another popular theory about DIA is that the airport's distinctive architecture and artwork are full of hidden symbols and messages that reveal the true purpose of the airport. The airport's design has been described as "satanic" by some conspiracy theorists, with the runways arranged in the shape of a swastika and a statue of a demon horse with glowing red eyes that some believe is meant to symbolize death and destruction.

While these theories may seem far-fetched, they have gained significant traction among certain groups of people, who point to a number of unusual features of the airport as evidence of a deeper conspiracy.

For example, some people have pointed out that the airport was significantly over budget and behind schedule when it was built, leading to speculation that the additional funds were used to construct the rumored underground bunker or city. Others have noted that the airport's artwork features a number of disturbing images, including a mural that depicts a gas-masked figure and dead children, and a sculpture of a gargantuan blue horse with glowing red eyes that fell on and killed its creator during construction.

Despite the persistent rumors and conspiracy theories surrounding DIA, there is no credible evidence to support the idea that the airport is anything other than a functional transportation hub. The airport's distinctive architecture and artwork are the result of a deliberate design scheme, and the airport has been subject to numerous safety and security inspections over the years.

While it is natural to be curious about the world around us and to question the motives and actions of those in power, it is important to approach these questions with a healthy dose of skepticism and critical thinking. Conspiracy theories can be both entertaining and potentially dangerous, as they can lead to a sense of paranoia and a mistrust of authority.

Ultimately, the mystery of the Denver International Airport is one that is likely to persist, as long as there are those who are convinced that there is more going on beneath the surface than meets the eye. However, for most people, the airport remains a functional and necessary part of the transportation infrastructure, with no evidence of any deeper conspiracy or hidden agenda.

96

The Secret of HAARP

The High Frequency Active Auroral Research Program (HAARP) is a research facility in Alaska that has been the subject of controversy and conspiracy theories for decades. The facility is designed to study the Earth's ionosphere and investigate the potential for using high-frequency radio waves for communication and other purposes. However, some people believe that HAARP is involved in more sinister activities, such as mind control and weather manipulation.

History of HAARP

The HAARP facility was built in the 1990s by the US Air Force, the US Navy, and the University of Alaska Fairbanks. Its purpose was to study the ionosphere, which is the upper layer of the Earth's atmosphere that is ionized by solar radiation. The ionosphere plays an important role in the propagation of radio waves and can have an impact on communication and navigation systems.

HAARP consists of an array of high-frequency antennas that can generate radio waves in the ionosphere. By studying how these waves interact with the ionosphere, researchers hope to gain a better understanding of this important region of the atmosphere and develop new technologies for communication and other applications.

Conspiracy Theories about HAARP

Despite the legitimate scientific research being conducted at HAARP, the facility has become the subject of numerous conspiracy theories and rumors. Some people believe that HAARP is involved in a range of secretive and nefarious activities, including:

1. Mind Control: Some conspiracy theorists believe that HAARP is capable of controlling people's minds through the use of electromagnetic waves.
2. Weather Control: Others believe that HAARP is capable of manipulating the weather, causing natural disasters such as hurricanes, earthquakes, and tornadoes.
3. Secret Military Experiments: Some believe that HAARP is being used by the military to conduct secret experiments, including the testing of new weapons and the creation of artificial earthquakes.
4. Connection to Alien Activity: Some people believe that HAARP is connected to extraterrestrial activity and that the facility is being used to communicate with alien civilizations.

Evidence against the Conspiracy Theories

Despite the persistence of these conspiracy theories, there is no credible evidence to support them. The scientific research being conducted at HAARP is legitimate and has been peer-reviewed by experts in the field. The facility has also been subject to numerous safety and environmental inspections, and there is no evidence to suggest that it is involved in any illegal or unethical activities.

Furthermore, the technology used at HAARP is not capable of mind control or weather manipulation. While it is true that radio waves can have an impact on the ionosphere, the effects are not significant enough to cause natural disasters or manipulate people's thoughts or behavior.

Conspiracy theories about HAARP have been debunked by numerous scientific organizations, including the American Association for the Advancement of Science and the Union of Concerned Scientists. These organizations have emphasized that the research being conducted at HAARP is legitimate and important for advancing our understanding of the Earth's atmosphere and improving communication and navigation systems.

The conspiracy theories surrounding HAARP are a reminder of the power of misinformation and the importance of critical thinking. While it is natural to be curious about the world around us and to question the motives of those in power, it is important to approach these questions with a healthy dose of skepticism and to seek out credible sources of information.

The research being conducted at HAARP is an important part of our scientific understanding of the Earth's atmosphere and has the potential to lead to important advances in communication and other fields. While conspiracy theories may be entertaining, they do not provide any real insight into the world around us and can distract from the real work being done by researchers and scientists.

Here are some facts about HAARP:

1. HAARP is located in Gakona, Alaska and consists of 180 antennas arranged in an array that covers an area of 40 acres.
2. The facility was originally built by the US Air Force and the US Navy in the 1990s, but is now owned by the University of Alaska Fairbanks.
3. The antennas at HAARP can generate radio waves with a frequency of up to 3.6 megahertz.
4. HAARP's primary objective is to study the ionosphere, which is the upper layer of the Earth's atmosphere that is ionized by solar radiation.
5. The ionosphere plays an important role in the propagation of radio waves, and by studying it, researchers hope to develop new technologies for communication and other applications.
6. HAARP has been used to study a variety of phenomena, including the effects of solar flares on the ionosphere and the properties of auroras.
7. The technology used at HAARP is not capable of mind control or weather manipulation, despite persistent conspiracy theories to the contrary.
8. HAARP has been subject to numerous safety and environmental inspections and has been found to be in compliance with all applicable regulations.

9. In recent years, the US government has reduced funding for HAARP, and the facility has been used primarily for scientific research conducted by the University of Alaska Fairbanks.
10. HAARP is one of several similar research facilities located around the world, including the European Incoherent Scatter Scientific Association (EISCAT) in Norway and the Arecibo Observatory in Puerto Rico.

97

The Truth Behind the Illuminati

The Illuminati is a secret society that has been the subject of conspiracy theories and speculation for centuries. The group is believed to be composed of some of the world's most powerful and influential people, and to have a global agenda that includes controlling world events and shaping the course of human history.

The origins of the Illuminati can be traced back to the 18th century, when a group of intellectuals in Bavaria, Germany formed a secret society called the Order of the Illuminati. The group was founded by Adam Weishaupt, a professor of law at the University of Ingolstadt, and was inspired by the ideals of the Enlightenment, such as reason, liberty, and equality.

The Order of the Illuminati was originally intended as a forum for discussion and debate among intellectuals, but it quickly became the subject of suspicion and controversy. The group was accused of plotting to overthrow the monarchy and the church, and was banned by the government in 1784.

Despite the ban, the influence of the Illuminati continued to spread, particularly among freemasons, who adopted many of the group's ideas and symbols. In the centuries since its founding, the Illuminati has become the subject of numerous conspiracy theories and rumors, with many people believing that the group is still active and working to achieve its goals of world domination.

However, the truth behind the Illuminati is far less sinister than the rumors would suggest. While the group certainly existed in the 18th century and had a significant impact on the intellectual and cultural landscape of the time, there is no evidence to suggest that the Illuminati has survived into the present day or that it is actively working to control world events.

Moreover, many of the ideas and symbols associated with the Illuminati, such as the pyramid and the all-seeing eye, have been taken out of context and misinterpreted by conspiracy theorists. These symbols have been used in a variety of contexts throughout history and do not necessarily have any connection to the Illuminati.

So why does the Illuminati continue to capture the imagination of so many people? One reason may be that the group represents a kind of mythical bogeyman, a shadowy organization that is believed to be behind everything from the assassination of JFK to the September 11th attacks.

Moreover, the idea of a secret society of powerful and influential people who are working behind the scenes to shape world events taps into a deep-seated human desire for order and control in a chaotic and unpredictable world. By attributing events to the machinations of a secret group, we can make sense of a world that often seems to defy explanation.

However, while the Illuminati may be a fascinating topic for conspiracy theorists and armchair detectives, it is important to remember that there is no evidence to suggest that the group still exists or that it is actively working to control world events. Moreover, the obsession with the Illuminati can distract us from the real issues that are shaping the world today, such as climate change, economic inequality, and political polarization.

The Illuminati is a complex and often misunderstood topic that has been the subject of speculation and conspiracy theories for centuries. While the group certainly had a significant impact on the intellectual and cultural landscape of the 18th century, there is no evidence to suggest that it still exists or that it is actively working to control world events. Rather than focusing on mythical bogeymen like the Illuminati, we should be focusing our attention on the real issues that are shaping the world today.

98

The Enigma of Area 51

Area 51 is a highly classified United States Air Force facility located in the Nevada desert. It has been the subject of much speculation and conspiracy theories, with many people believing that it is home to extraterrestrial technology and government secrets.

The history of Area 51 dates back to the 1950s, when the US government began testing advanced aircraft at the site. The area was chosen for its remote location and ideal testing conditions, including clear skies and minimal population.

Over the years, Area 51 has been the site of numerous high-profile aircraft development programs, including the U-2 spy plane, the SR-71 Blackbird, and the F-117 Nighthawk. These aircraft were designed and developed in secret, and their existence was not acknowledged by the government until many years later.

Despite the government's efforts to keep Area 51 secret, the facility has become the subject of intense scrutiny and speculation. Many people believe that the government is hiding extraterrestrial technology at the site, and that Area 51 is home to secret government projects that are too sensitive to be disclosed to the public.

One of the most persistent conspiracy theories about Area 51 is that the government is hiding evidence of extraterrestrial life. Some people believe that the facility is home to crashed UFOs and that the government is studying and reverse-engineering alien technology. There have also been reports of sightings of mysterious aircraft and lights in the skies above Area 51, further fueling speculation about the government's activities.

While the government has consistently denied the existence of extraterrestrial technology at Area 51, this has done little to quell the rumors and speculation. The secrecy surrounding the facility has only served to heighten suspicions and fuel conspiracy theories.

In recent years, some information about the government's activities at Area 51 has been declassified, shedding light on some of the facility's more mundane activities. For example, it has been revealed that the site was used to test new weapons systems and to develop new reconnaissance technology. However, many questions about Area 51 remain unanswered, and the government continues to maintain a high level of secrecy around the facility.

The secrecy surrounding Area 51 has led to a great deal of mistrust and suspicion among the public, with many people believing that the government is hiding something sinister at the site. Some have even gone so far as to suggest that the government is engaging in illegal activities, and that Area 51 is a symbol of the government's disregard for transparency and accountability.

In recent years, the government has taken steps to address some of the concerns surrounding Area 51. In 2013, the CIA officially acknowledged the existence of the facility, and released declassified documents that shed some light on the government's activities at the site. The move was seen as a step towards greater transparency and openness on the part of the government.

Despite these efforts, however, many questions about Area 51 remain unanswered, and the government's continued secrecy around the facility only serves to fuel conspiracy theories and speculation. Whether or not the truth about Area 51 will ever be fully revealed remains to be seen, but the facility will no doubt continue to capture the imagination of conspiracy theorists and skeptics alike for years to come.

10 facts about the enigma of Area 51:

1. Area 51 is a highly classified United States Air Force facility located in the Nevada desert.
2. The site has been shrouded in secrecy since its inception in the 1950s and is the subject of numerous conspiracy theories and rumors.
3. The facility is believed to have been used for the development and testing of advanced military aircraft and weapons systems, including the famous SR-71 Blackbird spy plane.
4. The government did not publicly acknowledge the existence of Area 51 until 2013, when it was declassified and the CIA officially acknowledged its existence.
5. Despite the declassification, many aspects of Area 51 remain highly classified and secretive.
6. The surrounding area of Area 51 is heavily guarded and patrolled, with security personnel known as "cammo dudes" keeping watch over the perimeter.
7. The secrecy surrounding Area 51 has fueled a range of conspiracy theories, including claims of extraterrestrial activity and the storage of captured alien spacecraft.
8. In 1989, a man named Bob Lazar claimed to have worked on extraterrestrial technology at a facility near Area 51 known as S-4. His claims have been heavily debated and remain unverified.
9. In recent years, the U.S. government has released a series of declassified documents related to Area 51, shedding some light on its operations and history.
10. Despite the continued speculation and rumors surrounding Area 51, the U.S. government maintains that the facility is simply used for military testing and training purposes.

99

The Power of the Military-Industrial Complex

The military-industrial complex (MIC) refers to the close relationship between the military and defense industry, which has been a driving force behind the growth of the defense industry and the development of new military technologies. This complex is a powerful and influential force in modern society, and has far-reaching implications for national security, economic growth, and international relations.

The origins of the military-industrial complex can be traced back to the post-World War II era, when the United States emerged as a global superpower and faced new threats from the Soviet Union and other hostile powers. In response to these threats, the US government invested heavily in defense spending, which fueled the growth of the defense industry.

The relationship between the military and defense industry is complex and multifaceted. On one hand, the military relies on the defense industry to develop and produce new weapons systems and technologies, which are essential for maintaining military superiority and ensuring national security. The defense industry, in turn, relies on the military for funding and contracts, which provide the financial resources needed to develop and produce new technologies.

At the same time, the close relationship between the military and defense industry has raised concerns about the influence of corporate interests on national security policy, and the potential for conflicts of interest and corruption. Critics argue that the military-industrial complex has become a powerful lobby that exerts undue influence on government policy, and that defense spending has become a mechanism for corporate welfare rather than a means of promoting national security.

Despite these concerns, the military-industrial complex remains a key driver of innovation and technological progress in the defense industry. The close collaboration between the military and defense industry has led to the development of advanced weapons systems, such as drones, stealth technology, and precision-guided munitions. These technologies have revolutionized modern warfare, and have enabled the military to achieve unprecedented levels of precision and effectiveness on the battlefield.

The military-industrial complex also has significant economic implications. Defense spending is a major source of employment and economic activity in many parts of the United States, particularly in areas with a high concentration of defense contractors and military installations. The defense industry is a major employer and a key source of innovation and technological progress, and has helped to drive economic growth and prosperity in many parts of the country.

At the same time, the military-industrial complex has been criticized for contributing to economic inequality and social dislocation. Defense spending is often concentrated in a small number of regions and industries, and the economic benefits of defense spending are not distributed evenly across society. Critics argue that the military-industrial complex has contributed to the hollowing out of many manufacturing industries in the United States, and has led to the concentration of wealth and economic power in the hands of a few large corporations.

The military-industrial complex also has significant implications for international relations and global security. The close relationship between the military and defense industry has led to the development of new weapons systems and technologies that have the potential to alter the balance of power between nations. The proliferation of advanced military technologies has raised concerns about the potential for arms races and increased instability, particularly in regions with a history of conflict and tension.

Despite these concerns, the military-industrial complex remains a powerful and influential force in modern society. The close collaboration between the military and defense industry has led to the development of new technologies and capabilities that have helped to ensure national security and promote economic growth. At the same time, the military-industrial complex has raised important questions about the role of corporate interests in national security policy, and the potential for conflicts of interest and corruption.

As the United States faces new threats and challenges in the 21st century, the role of the military-industrial complex is likely to remain a subject of intense debate and scrutiny. While the close relationship between the military and defense industry has contributed to significant advances in national security and technological progress, it has also raised concerns about the potential for corporate interests to influence government policy, and the need for greater transparency and accountability in defense spending and procurement.

One of the key challenges facing the military-industrial complex is the need to balance national security and economic growth with the need to ensure accountability and transparency in defense spending. The defense industry is a key driver of economic growth and technological progress, but it must also be subject to rigorous oversight and scrutiny to ensure that it serves the national interest and is not used for the benefit of narrow corporate interests.

Another challenge is the need to address the growing threat posed by cyber warfare and other non-traditional forms of conflict. As the nature of warfare changes, the military-industrial complex must adapt to new threats and challenges, and develop new technologies and capabilities to address emerging threats.

In recent years, there has been growing concern about the impact of the military-industrial complex on democratic institutions and values. Critics argue that the close relationship between the military and defense industry has led to a culture of secrecy and secrecy that undermines public trust in government and undermines democratic values such as transparency and accountability.

To address these concerns, there have been calls for greater transparency and oversight in defense spending and procurement, and for greater account-ability and public engagement in national security policy. Some advocates have also called for a renewed focus on diplomacy and international coopera-tion as a means of addressing global security challenges, rather than relying solely on military force and the defense industry.

The military-industrial complex is a powerful and influential force in modern society, with significant implications for national security, eco-nomic growth, and international relations. While the close relationship between the military and defense industry has contributed to significant advances in technology and national security, it has also raised important questions about the role of corporate interests in government policy and the need for greater transparency and accountability in defense spending and procurement. As the United States faces new threats and challenges in the 21st century, the role of the military-industrial complex will continue to be a subject of intense debate and scrutiny, and will require careful management and oversight to ensure that it serves the national interest and promotes the values of transparency, accountability, and democracy.

Here are the 10 most powerful militaries in the world:

1. United States: The United States has the largest military budget in the world and the most technologically advanced military.
2. Russia: Russia has a large and powerful military with a significant nuclear arsenal.
3. China: China has the largest standing army in the world and is rapidly modernizing its military capabilities.
4. India: India has a large and capable military with significant air and naval power.
5. France: France has a highly professional and well-equipped military

with significant capabilities in special operations.

6. United Kingdom: The UK has a modern and well-equipped military with significant air and naval power.

7. Japan: Japan has a technologically advanced military and is investing in developing its defense capabilities.

8. Turkey: Turkey has a large and well-trained military with significant capabilities in armored warfare.

9. Germany: Germany has a highly professional and well-trained military with significant air and naval power.

10. South Korea: South Korea has a well-trained and well-equipped military with significant capabilities in special operations and cyber warfare.

20 facts about the power of the military-industrial complex:

1. The term "military-industrial complex" was first coined by U.S. President Dwight D. Eisenhower in his farewell address to the nation in 1961.

2. The military-industrial complex refers to the close relationship between the government, the military, and the defense industry.

3. The complex is fueled by government spending on defense and military-related projects, which in turn drives innovation and technological advances in the defense industry.

4. The United States spends more on defense than any other country in the world, with a budget of over $700 billion in 2020.

5. The military-industrial complex is estimated to employ over 3 million people in the United States alone.

6. The complex has been criticized for promoting a culture of war and

perpetuating conflicts around the world.

7. The complex has also been accused of driving government policy towards aggressive foreign interventions, including wars in Iraq and Afghanistan.

8. The military-industrial complex has a significant influence on political lobbying and campaign contributions in the United States.

9. The complex has been accused of fostering corruption and conflict of interest among government officials and defense industry executives.

10. The military-industrial complex has been linked to the rise of private military contractors, which have been criticized for their lack of accountability and human rights abuses.

11. The complex has been accused of promoting a culture of fear and paranoia in the United States, particularly in the wake of the 9/11 attacks.

12. The military-industrial complex has also been linked to the surveillance and security industries, with the development of advanced technologies for tracking and monitoring individuals.

13. The complex has been criticized for diverting resources away from social programs and domestic issues, such as healthcare and education.

14. The military-industrial complex has a significant impact on the global arms trade, with the United States being the largest exporter of arms in the world.

15. The complex has been accused of promoting a militarized approach to international relations, rather than diplomacy and conflict resolution.

16. The military-industrial complex has been linked to the development of nuclear weapons and other weapons of mass destruction.

17. The complex has been criticized for its environmental impact, with defense-related activities contributing to pollution and climate change.

18. The military-industrial complex has been linked to the growth of the prison-industrial complex, with the development of private prisons and detention facilities.

19. The complex has been accused of perpetuating racism and inequality, with a disproportionate number of people of color and low-income

individuals being impacted by defense-related activities.

20. The military-industrial complex remains a controversial and highly influential force in global politics and economics, with its impact felt in almost every aspect of modern society.

100

The Secret of the Bilderberg Group

The Bilderberg Group is a secretive and exclusive organization made up of influential political leaders, business executives, and intellectuals from around the world. Its annual meetings, which are closed to the public and the press, have been the subject of intense speculation and conspiracy theories.

The group was founded in 1954 by Prince Bernhard of the Netherlands, who was concerned about the growing anti-American sentiment in Europe following World War II. He invited a group of European and American leaders to a conference in the Netherlands to discuss ways to promote closer ties between Europe and the United States.

Since then, the group has met annually in different locations around the world, with the attendance of around 130 participants from North America and Europe. The meetings are closed to the public and the press, and participants are sworn to secrecy about what is discussed.

The Bilderberg Group is often accused of being a secretive cabal that seeks to control world events and promote a globalist agenda. Conspiracy theories have linked the group to everything from the New World Order to the Illuminati, and its annual meetings are a favorite topic of fringe groups and conspiracy theorists.

However, the group maintains that its meetings are simply an opportunity for influential leaders to discuss important issues and exchange ideas in a private setting. According to the group's website, the meetings are "designed to foster dialogue between Europe and North America" and "to contribute to better understanding of the complex forces and major trends affecting Western nations in the difficult years ahead."

Despite the group's claims of transparency, its meetings have been the subject of intense speculation and scrutiny. In 2013, for example, protestors gathered outside the Bilderberg meeting in the UK, accusing the group of promoting a globalist agenda and conspiring to undermine democracy.

The Bilderberg Group has been accused of promoting a range of controversial policies and agendas, including globalization, free trade, and military interventionism. Critics argue that the group's exclusive membership and secretive meetings promote elitism and undermine democracy by excluding the public and the press from important policy discussions.

Supporters of the group, however, argue that its meetings provide an important forum for influential leaders to discuss important issues and exchange ideas. They point to the group's role in promoting closer ties between Europe and North America, and its contributions to important policy debates such as the European Union and the global economy.

While the exact nature and purpose of the Bilderberg Group remain shrouded in mystery, it is clear that the group plays a significant role in shaping the world's political and economic landscape. Its annual meetings bring together some of the world's most influential leaders and thinkers, and provide a forum for discussing important issues and exchanging ideas.

As the world becomes increasingly interconnected and complex, the need for such a forum may become even more pressing. However, the group's secretive nature and exclusive membership raise important questions about accountability, transparency, and democracy. As the world's leaders continue to grapple with these challenges, the Bilderberg Group is likely to remain a controversial and enigmatic presence in global politics.

10 facts about the Bilderberg Group:

1. The Bilderberg Group is named after the Hotel de Bilderberg in the Netherlands, where its first meeting was held in 1954.

2. The group's annual meetings are attended by around 130 participants from North America and Europe, including political leaders, business executives, and intellectuals.

3. The meetings are closed to the public and the press, and participants are sworn to secrecy about what is discussed.

4. The group's website describes its meetings as "designed to foster dialogue between Europe and North America" and "to contribute to better understanding of the complex forces and major trends affecting Western nations in the difficult years ahead."

5. Some of the most influential political leaders and thinkers in the world have attended Bilderberg meetings, including Margaret Thatcher, Bill Clinton, Angela Merkel, Henry Kissinger, and Tony Blair.

6. The group has been accused of promoting a range of controversial policies and agendas, including globalization, free trade, and military interventionism.

7. The Bilderberg Group has been the subject of intense speculation and conspiracy theories, with some accusing the group of promoting a globalist agenda and conspiring to undermine democracy.

8. The group's influence and power have been questioned by some, with critics arguing that its secretive nature and exclusive membership undermine democracy and accountability.

9. Despite its controversies and criticisms, the Bilderberg Group has been credited with helping to promote closer ties between Europe and North America, and with contributing to important policy debates such as the European Union and the global economy.

10. The group continues to hold annual meetings, and its influence and role in shaping global politics and economics remain a subject of intense debate and speculation.

www.ingramcontent.com/pod-product-compliance
Lightning Source LLC
Chambersburg PA
CBHW071131220526
45467CB00015B/846